FAITH THEOLOGY AND IMAGINATION

TO JAN

FAITH THEOLOGY AND IMAGINATION

JOHN McINTYRE

Professor Emeritus of Divinity
University of Edinburgh

1987
The Handsel Press

Published by
The Handsel Press Ltd.
33 Montgomery Street, Edinburgh

© 1987 John McIntyre

All rights reserved. No part of this publication may be reproduced, stored in a retrieval system, or transmitted, in any form or by any means, electronic, mechanical, photocopying, recording or otherwise, without the prior permission of The Handsel Press Ltd.

British Library Cataloguing in Publication Data

McIntyre, J.
   Faith theology and imagination.
   1. Religion — Philosophy
   I. Title
   200'.1   BL51

ISBN 0-905312-65-1

Printed in Great Britain by Bell and Bain Ltd., Glasgow

FOREWORD

The present study has had a somewhat extended life-span. It had its beginnings in the McCahan Lecture delivered at what was then known as Assembly's College, Belfast, on 25 May 1962, and published in that form soon after in *The Expository Times*. After following up certain other interests in the doctrine of God and Christology, I returned to the subject of imagination in relation to faith and theology in four lectures given in St Andrew's College within the University of Sydney, on the occasion of the centenary of the College in 1970. But it was not until the session of 1977-78, that I was free to extend these four lectures in a variety of different directions. The opportunity came through sabbatical leave, the first half of which, in 1977, was spent in the Philosophy Department of the University of New England, Armidale, New South Wales, where Professor Dick Franklin kindly encouraged me to provide a course on the subject of imagination as treated by a series of philosophical writers, from Plato to Sartre. Some of that material appears in what follows, though in a much reduced form. The second half of that sabbatical leave, in 1978, was spent at Princeton Theological Seminary, through the good offices of the then President, Dr J I McCord; and it was there that I was able to develop the more theological side of the study. The invitation from Principal Adam Neville to give the Margaret Harris Lectures on Religion in November 1984 encouraged me to re-think the material, and to prepare it in a final form for publication. I should like now to express my very sincere thanks to all the persons whom I have named who by their generosity have made it possible for me to pursue my interest in a subject which has been greatly neglected in traditional theology, and which, as I shall hope to show, is so central to the whole theological enterprise. It would, also, be remiss of me to fail to thank the students in three continents who by their questioning and discussion have maintained my own interest in these issues, or Mrs Joan Hunter, the Department Secretary, who has so uncomplainingly produced readable manuscript from poor originals. Acknowledgment is very gratefully made to the Revd Dr C S Rodd, Editor, for permission to use themes from the McMahan Lecture printed in *The Expository Times* in 1962; to the Revd Dr John S Bowden, Managing Director, SCM Press, for permission to use the summary form of my analytic of imagination, appearing in the article, 'Imagination', by myself in *The New Dictionary of Christian Theology*; and to Miss Judith Longman, General Editor, SPCK, for similar permission to use the summary form of my analytic of images, contained in my article on 'Imagination' in *The Dictionary of Pastoral Theology*.

I should like to take this opportunity to express a special word of appreciation to Dr Douglas Grant, Chairman of Directors of the Handsel Press, for his supervision of the publication of this manuscript. For he it was who, when Managing Director of Oliver and Boyd, launched me on the way of publication with *St Anselm and His Critics* — over thirty years ago.

CONTENTS

		Page
1.	Introduction with Reference to *A Dish of Orts*	1
2.	The Parabolic Imagination	19
3.	Imagination as a Theological Category	41
4.	Imagination: The Ethical Dimension	65
5.	Imagination: The Philosophical Dimension	89
6.	Imagination: Its Place in Methodology and Epistemology	127
7.	An Analytic of Imagination and Images	159

CHAPTER 1

INTRODUCTION
WITH REFERENCE TO *A DISH OF ORTS*

It is now almost fifty years since John Baillie expressed a view on the place of imagination within historical Christianity which provides the setting for the present study. It is a view which he did not subsequently develop, nor did it seem to form a major theme in his thinking. Yet, uttered when it was, it has come to have a certain prophetic quality, in view of the considerable interest which the subject has created in recent years. (Alastair Hannay, *Mental Images: A Defence*, 1971; Ray Hart, *Unfinished Man and the Imagination*, 1974; Mary Warnock, *Imagination*, 1976; David Tracy, *The Analogical Imagination*, 1981). What Baillie said was, 'I have long been of the opinion that the part played by the imagination in the soul's dealings with God, though it has always been understood by those skilled in the practice of the Christian cure of souls, has never been given proper place in Christian theology, which has been too much ruled by intellectualist preconceptions' (*Our Knowledge of God*, p.77).

Baillie's remark is singularly concentrated, for he seems to be implying four distinct if related points. Not only is he deploring the extent to which our theological tradition has been over-conceptualised, a fact as true of the protestant as of the catholic systems, but he is also discovering a lack in the quality of the Christian life as it has been classically portrayed. Further, he is suggesting a positive task of constructing a theology in which a role might be assigned to imagination; or more precisely, the role which those who have a cure of souls have already discovered within Christian faith and life might be taken as indicative of its presence also in the manner of our thought about that life and faith in our theology and ethics, as well as in our religious epistemology and theory of homiletics and the devotional life.

Baillie has therefore given us what might be regarded as the long-term objective of our study. But such an objective is not of great value unless the road to it is charted in the middle distance. So we now turn to medium-term objectives which we shall be attempting to achieve. First, since the concept of imagination has held a quite central place in western thought for some two and a half thousand years, we shall be wise to recognise that it has very probably been used in many different senses. It exists now in our vocabulary, carrying many of the overtones of other days and other cultures. Like other words, it encapsulates its past, a past full of interest to us in our present task. In recognising the pre-history of the term from Plato and Aristotle, the neo-Platonists and St Thomas, Hume, Kant and Coleridge, to Sartre, Ryle and Warnock, we are not falling into the fallacy that the philosophical use of the concept must be taken as determinative of any theological application that we intend to make. In fact our minds should be sufficiently open to expect the theological reference of the concept to bring its own penetrating illumination. Ignoring these other usages, however, would be not only unscholarly if not even irresponsible; it would lead to a grave impoverishment in our understanding of it.

A second medium-term objective of the study follows immediately from the first, namely, that of achieving some definition of the notion of imagination; or less ambitiously, to draw out of the wide variety of the uses of the notion, those elements which should be regarded as essential to it. As has been suggested above, it has been understood in many ways across the centuries, and so, if we employ it at all, we are inevitably involved in value-judgments about what appear as its central and inalienable features, and those which have to be taken as expendable, or maybe on occasion as extravagant. Difficult and at times controversial as this part of the task may prove, it is nevertheless not to be shirked for that reason. Only by aiming at such clarification and systematic presentation, can we hope to discern and describe the important part which imagination has played and may still play in Christian life and thought.

A third objective unfolds itself when we turn from the philosophical and cultural milieux in which the notion has appeared in western thought, to enquire into its place in the thought of the Bible. At a first glance, and for reasons to which we shall soon turn, it might be feared that we are committing the solecism of imposing upon the Bible, alien categories which distort and falsify its meaning. Maybe it is a measure of our own obsession with conceptualisation that we should express such a fear; for *prima facie* the stories, the poetry, the myths, the parables and the apocalypses of the Bible would seem to be more amenable to description in terms of imagination and the images which are its media, than to categorisation according to Aristotelian or other logical structures. In the event it may also be possible that the biblical material will have its own special contribution to make to our understanding of what imagination is, of how it works and of the place which it occupies in our mental and spiritual life.

Tracing the role of the imagination and of imagery in the biblical material leads naturally on to a further and fourth objective, namely, to explore the part that it might be thought to have in theology itself, or more precisely, in the subject-matter which theology delineates. Here we have to admit that we are entering much more intractable territory, remembering Baillie's strictures upon the over-intellectualisation of theology. Nevertheless, despite such anticipated difficulty, we can not allow it to impede our serious intention to ensure that our theology reflects with accuracy the biblical understanding of God and of all his works. In fact, as we pursue the concept of imagination into theology itself, we may be looking for it in a variety of places. For example, we could be seeking for it in the actual content of theology, attempting to discover whether in our description of God, or, say, in our presentation of the work or the person and teaching of Jesus Christ, there are fresh insights to be derived from the idea of imagination which have not been used extensively in theology before, or perhaps not at all, and which are nevertheless permissible and valid. As the history of the concept in philosophy shows, it is bound up with the process in which we get to know the world around us and the people with whom we share this world. We shall therefore be encouraged to look again at the question of the nature of our knowledge of God in the light of that history of epistemology, and to enquire afresh about the nature of faith. Remembering also Baillie's reference to the place of imagination in the activities of those who practise the cure of souls, we shall wish to acknowledge the place of imagination in the devotional life, the world of liturgy and worship, the whole sphere of communication, as well as of Christian worship.

At this point we shall have to gather into a single objective, our fifth, a consideration of the nature of images and imagery, of their place and status within religion and theology, and of their validity in the wide and heterogeneous purposes which they serve. There is a sense in which the investigation concerning the nature of images is to be regarded as independent of the previously mentioned examination of imagination; for while some would seriously contest the propriety of including imagination in any reputable theological enquiry, they would find it very difficult to exclude a discussion of images, particularly as they are frequently now considered in the context of such notions as models and paradigms. In other words, while it may traditionally have been thought that imagination and images are exact correlates, and that imagination is about images, it would now appear that they have a certain independence of one another. On the one hand, imagination may exercise itself on other subjects than images, while on the other, images would seem to have an active role, and not simply be the passive creations of the imagination.

If imagination should then turn out to be as prevalent in theology as the above description of our objectives would suggest, the question could then be asked whether we have not indicated yet another objective which is concerned with the very nature of how we do theology. That, I would propose, is our sixth objective, not to invent some new theology which is designed to replace the old, but rather, by using this concept of imagination, to work over much of the familiar theological material, to view it from a different angle, in the hope that we would gain fresh understanding of our faith, of the nature of God, in his triune being and activity, on how we are to fulfil

INTRODUCTION 5

our Christian duty and practise prayer, and so on. Interpreted in these terms our enterprise becomes a question of the kind of theology we want to do and of how we want to do it, a question of the kind of Gospel we intend to preach and of the nature of the response that we expect from those who hear it, and a question of how we endeavour to relate to our fellows, to society, to nature and to the future.

Having stated these objectives, we may well be thought to be somewhat over-ambitious, not to say totally insensitive to the misgivings which many would feel to be a thoroughly misconceived enterprise. The reasons which might be offered for such misgivings would be very varied indeed. Rather, therefore, than falter at the starting-post, let us take some time to consider these objections:

First of all, it might be possible to adopt a literary tack, and draw attention to the fact that the single English word 'imagination' has been employed in the Authorised Version of the Bible to translate three different Hebrew words in the Old Testament — יֵצֶר *yetser* (e.g., Gen 6.5; 8.21; Deut 31.21), שְׁרִירוּת *sheriruth* (e.g., Deut 9.19; Jer 3.17; 7.24; 9.14; 11.8; 13.10; 16.12; 18.12; 23.17), מַחֲשֶׁבֶת *machashebeth* (e.g., Prov 6.18; Lam 3.60f); *and* three different Greek words in the New Testament — διαλογισμός *dialogismos* (Rom 1.21); διάνοια *dianoia* (Lk 1.51); λογισμός *logismos* (2 Cor 10.5). The interesting point about all of these references is that they are pejorative, uniformly implying that imagination is unacceptable to God in all its machinations, and therefore to be eliminated from the minds and thoughts of his servants. Now the translations of the six different Hebrew or Greek words by the one word 'imagination' may well have been faulty in most of the cases noted, and the fact that they have all been differently translated in almost all of the many modern versions would seem to suggest that they were. Nevertheless, the universality of use of the Authorised Version throughout the English-speaking nations for over three and a half centuries, together with its immense influence not just upon everyday speech but also upon the thought of everyday people, has ensured almost universal suspicion of the concept itself. It is not difficult therefore to see why there is no readiness to make a place for it in theological thought. The modern translations by their above-mentioned abandonment of the term in the passages referred to, have certainly eased the passage towards the introduction of the notion of 'imagination' into theology in a positive and constructive role.

Next, the objectors might wish to quote Calvin (*Institutio*, 1.xi.i) to us: 'God rejects without exception all shapes and pictures, and other symbols by which the superstitious imagine they can bring God near to them. These images defile and insult the majesty of God'. Calvin does have other things to say about symbols, which are a form of image, particularly about the one true image, the Son, who is the express image of the Father; and about those symbols, bread and wine, which when received in faith become the body and blood of our Lord. It would be wrong therefore to identify Calvin with any wholesale rejection of symbols and images, some of which clearly have a place within true religion; and a fuller discussion of the subject would take us away from our present subject. Nevertheless, there has entered into the popular protestant tradition, from Calvinism, if not from Calvin himself, a certain sequence of thought, together with its emotional concomitants which might be regarded as a kind of Rake's Progress. It runs as follows:

Stage one is the affirmation of a state of iconolatry in the worship, so it is thought, of certain denominations within Christianity. Such iconolatry is roundly condemned not simply for the two reasons given by Calvin, that images are wrongly thought to draw the transcendant and wholly other God within the manipulative control of his creatures, and that because of their inadequacy to represent such a God in visual form, images are an affront to his majesty. A still more serious reason for the condemnation of iconolatry is that it comes to identify God with the image, which as the history of Israel testifies is the final blasphemy.

Once the state of iconolatry has been established and identified, then stage two of the Rake's Progress follows, namely, iconoclasm, which involves the destruction of the images for the reasons involved in the rejection of iconolatry, and which has received so much notoriety in some histories of the Reformation. In Scotland the apologia is often presented that much of the damage done to pre-reformation churches derived not so much from the image-smashing tendencies of the reformers, as from scarcity of funds, due to the secularisation of ecclesiastical property, to maintain the fabric. There was, however, just enough of the theological component in the rejection of images — even when it was not accompanied by the large-scale vandalism of iconoclasm — to discourage any wholesale re-introduction of images into the protestant tradition once a case for the place of some form of images in worship had been formulated.

So stage three is reached, namely, iconophobia, the fear that by having any images at all in religion, we place ourselves on the beginning of the slippery slope which has but one ending, the blasphemy of iconolatry. This fear is in turn fed by the suspicion that if our faith requires the artificial aids of images it must be something less than genuine trust in the Lord Jesus Christ; we must be related to Christ by something other than *sola fide*. It is not unusual at this point to blame calvinism for the iconophobic strain in the protestant tradition in subsequent centuries, and for an alleged consequent totally inhibiting influence upon the arts. The criticism is invalid, because historically inaccurate, for there is ample evidence of much artistic activity, in poetry, prose, painting and music, in many of the countries affected by the calvinistic form of the Reformation. In Scotland at the present time, there is a considerable amount of poetic expression which rather paradoxically derives its inspiration from the assumption that calvinism particularly in Knox's presentation of it, killed all joy and love of beauty, as well as all appreciation of nature and ordinary human nature. In fact — to digress briefly — it was in Scotland more likely to have been eighteenth century puritanism which spread the blight, and could well have exercised the same kind of influence in other countries which shared the calvinistic tradition. Often, too, we have to distinguish Calvin from calvinism and not charge the master with the imperfections and obsessions of his over-enthusiastic supporters. One very positive consequence of the admittedly iconophobic strain in the protestant tradition is a realisation, particularly in ecclesiastical architecture, that beauty need not be ornate, and that there is beauty also in simplicity, especially when the physical simplicity is matched by a like simplicity in liturgy.

These three previous stages — of rejection of iconolatry, iconoclasm and iconophobia — have re-inforced a process which has been characteristic of theology since at least the third century, sufficiently to have set a fourth stage to our Rake's Progress, namely, what I shall call, if I may be allowed a neologism, aniconastic thinking, or imageless thought, as being the norm of theological reflection and articulation. In other words, to be seeking ways in which we may re-constitute and re-establish imagination within religion and theology, we are running counter to the theological tide of the ages, which has all along been moving in the direction of continuous sterilisation of theology and religion of all the images with which it is thought to be encumbered and incapacitated, and the re-formulation of what the images were attempting to convey, but now in appropriate and intellectually acceptable conceptualised forms. Looked at from this perspective, the demythologisation enterprise of the past forty and more years, so often hailed in that time as the great new event of the century, can be seen as the modern form of a process which has gone on now for centuries in theology, and is not really very novel in the least. What originality there has been has lain partly in the quite explicit form which the enterprise has taken — previously it had been done without open acknowledgment or confession; and partly, in the contrived form which the re-formulation of the demythologised Gospel had taken, through the aid of existentialist philosophy. But some of the most notable expressions of aniconastic thought are to be found in such classical documents of the church as the Nicaeo-constantinopolitan and the Chalcedonian Creeds or the Westminster Confession, where we have such words as *hypostasis*, *physis*, predestination, regeneration, sanctification, to mention but a few to which it would be difficult to attach images.

It has to be admitted, therefore, that to enquire concerning the role of imagination, and by implication, of images in the Christian faith and theology, is to fly in the face of a tradition which is as old as theology itself. But we are prepared to make the effort and take the risk in the conviction that these areas of theological thought are due now for re-examination, based as they are upon what the Bible in the first place and the events of the Incarnation have to teach us both about our faith and about how we ought to think when we reflect upon it and endeavour to communicate it to others. The objectors to whom we are at present listening may go off on another tack, and quote the bad press which imagination has had at the hands of some of the outstanding thinkers of modern times. Hegel wrote, 'Theism in all its forms is an imaginative distortion of final truth' — a judgment, which if taken seriously and literally would commit its supporters to such an identification of theism with imaginative thought as to require the excision of both of them from all serious pursuit of the truth. For our part we seek a place for imagination in theology — but scarcely at such a price for either of these activities! Feuerbach, not altogether unpredictably, has a similar denigratory comment to make, 'What has been called the mythopoeic function of imagination creates the object of religion, which in that act stands revealed as delusion'. Once again the inter-penetration of imagination and religion yields a result which is totally unacceptable to the philosophical mind. Freud, without actually using the word 'imagination' perpetuates this same association of imagination with religion which results in a mythology which has no subject in the real world, but is only an inner construct with significance to the psychiatrist. Thus, 'I believe that a large portion of the mythological conception of the world which reaches far into most modern religions is nothing but psychology projected into the outer world. The dim projection of psychic factors and relations of the unconscious was taken as a model in the construction of a transcendental reality which is destined to be changed back again by science into the psychology of the unconscious' (*Psychopathology of Everyday Life*, ET, Benn, London, 1960, p.217). It is rather interesting to see how the demythologisers, whether of a theological or a secular variety, share the same dismissive assessment of imagination when it is in association with religion. What is not clear is whether it is religion which employs the imagination to create its delusions and misrepresentations of the truth, or whether it is the imagination which corrupts what would be otherwise a valid intellectual activity. Either way, the alleged mutual involvement of religion with imagination in a conspiracy to distort known truth or to invent patent error places under dire suspicion anyone who is moved to suggest that the relationship might be honourable or even reciprocally beneficial and enlightening. However no apologia either for the Christian faith or imagination will be offered at this stage; there is a sense in which this whole study is itself such an apologia, insofar as it shows how closely the two are related to one another. We certainly will not allow ourselves to be stampeded away from our central purpose of considering Christianity in its imaginative dimension because some expressions of the alleged pathological nature of religion can be shown to derive from origins of fantasy. But we have registered the warning that the term 'imagination' does have rather a chequered case-history.

Finally, serious misgiving about the whole enterprise might be felt in another quarter. The notion of revelation to which most Christian theologies are committed is sometimes presented as if it implied that there was either no participation by the believer beyond passive reception of a subject immediately given; or, if the situation was thought to require participation, it was described as faith, given by the Spirit, and allowing for no contribution from the human side. Viewed from such a perspective, the admission of imagination to any analysis of revelation and our response to it amounts to an undue weighting of the human element, opening the door even to the possibility that the human mind or heart might distort the pure given-ness of revelation. At a deeper level, but still within the context of consideration of revelation and our apprehension of it, it might be said that we should be very careful in defining the place which imagination and images may have within such apprehension, in view of the 'conceptual character of the knowledge of God' as it 'arises out of His self-disclosure in His Word' (Professor T. F. Torrance, *Theological Science*, Oxford. 1969, p.14). Professor Torrance is not advocating a 'radical cleavage' between concept and image — in fact he deplores the consequences of such a cleavage as it has been introduced into modern thought by Cartesian dualism. But in criticism of Austin Farrer who claims that 'we cannot by-pass the images to seize an imageless truth' (A. M. Farrer, *The Glass of Vision*, p.110), while recognising that 'in the biblical tradition image and word belong together' (*op. cit.*, p.20),

INTRODUCTION 11

Professor Torrance nevertheless holds that we are not permitted to have in our thought any 'imaginative or pictorial representation' of God (*ibid.*). Perhaps I may be allowed two comments here. First, as I have already indicated, following upon the short quotation from John Baillie mentioned earlier, I would be somewhat uneasy about an outright preference for concepts as against images at an early stage of an epistemological discussion on the nature of our knowledge of God. Farrer has admittedly over-stated his case in suggesting that God himself is 'imageable'. So I would agree with Professor Torrance when he quotes with approval Farrer's view that we cannot discern what the images signify except through the images. But in this respect, images are not greatly different from concepts which seem to perform a similar role in the process in which we apprehend God. We can not apprehend God except in terms of the conceptual structures which are the modes of rationality set up between God and ourselves in his self-communication to us. Images may also be 'open' or 'closed' as are concepts, and so have a flexibility not dissimilar to concepts. Secondly, I would be uneasy, at this early stage of our discussion, with any suggestion that images had to be 'pictorial' or 'representational', and I am not convinced that Farrer is necessarily claiming that they are either. If we make him that concession, and it may not in fact be a concession, then he may be exculpated from the charge of idolatry. We could then still disagree with him about the source of the imagery employed in statements about God, and depart from his use of the notion of natural knowledge, as a norm for interpreting revelation. Thirdly, I am however greatly encouraged by Professor Torrance's contention (*op. cit.*, p.20) that there is in the biblical revelation a place for a certain type of images which 'set the stage, as it were', and through which God is revealed in his Word. Such images are not, of course, to be identified with the word which accompanies them, far less with the Word to which they point. Nevertheless, it is not necessary to go on to say that therefore the images are somehow expendable, or that we do not apprehend God in terms of the images. It is not so much a matter of 'looking past' the images to the reality which they signify or indicate, as 'looking through' the images to that reality. To take this matter any farther is, however, to trespass upon a theme, namely, the epistemological role of images and imagination, which will be occupying our attention later. The case against any possible acknowledgement of a constructive place for imagination within religion and theology might therefore appear convincing. By some imagination is seen as the medium which diminishes the real character of religion; it distorts the relation in which religion stands to reality, misrepresenting the nature of God with whom its practitioners have to do. So it must well seem to be a well-nigh impossible task to rehabilitate and re-constitute imagination so that it may have not only a positively enlightening role but also a validly useful one in enabling us to understand better the subject-matter of theology. Therefore, if I were to sum up the state of play as it was when I first became interested in this question some twenty or more years ago, I would have to say that there was little hope of reversing a process which seemed to have the weight of centuries, and every canonical, and perhaps a few heretical creeds in its favour. How has change in this situation been effected? What is the dimension of the change — the change, that is, from anathematising and ostracising imagination from any appropriate place within all religious thought and activity?

Though there have been several books written within the field of imagination and theology, as referred to in the opening paragraph of this study (p.2), for me the change began when I came upon a little known essay by George MacDonald. George MacDonald has, of course, become a key-figure in the recent extensive research devoted to the work of C. S. Lewis — quite understandably, because as early as February 1916, Lewis acknowledged that the reading of MacDonald's 'Faerie Romance', *Phantastes*, had left the deepest and most enduring impression on his literary and spiritual life. So much so that thirty years later, in the introduction to a selection from his works (according to R. L. Green and W. Hooper, *C. S. Lewis*, Collins, London, 1974, pp.45f.), Lewis wrote, 'I have never concealed the fact that I regard MacDonald as my master; that whole book (i.e. *Phantastes*) had about it a sort of cool morning innocence; and also, quite unmistakeably, a certain quality of Death, *good*

INTRODUCTION

Death. What it did was to convert, even baptise (that was where the death came in) my imagination'. Though *Phantastes* did not of itself turn Lewis to Christianity, there is no denying that the baptism of his imagination into a new form for Romanticism was the beginning of a process which culminated in the Narnian series. Of great importance for anyone involved in Lewis's literary and spiritual development, is the writing of MacDonald's which we are now going to consider.

This same essay whose content is so germane to our purpose is entitled 'The Imagination: Its Function and its Culture', itself written in 1867, but not published until 1907 as the opening essay in a book entitled *A Dish of Orts*. What, the reader may well be forgiven for asking, are 'orts'? In reply, I might say, show me someone who knows what 'orts' are, and I will show you someone who counts cross-word solving among his or her pastimes. 'Orts' are left-overs; and a less imaginative author would have entitled the book *Collected Essays*. So I propose to introduce this study of the place of imagination in religion and theology by means of a short commentary upon this little known work, for it will provide us, in the event, with a ground-plan for much of our subsequent discussion. Somewhat of a classic in its own right, it failed apparently to produce any reactions in either theological or literary circles. Yet it will prove remarkably pertinent to our purpose in touching upon some of the philosophical themes that will concern us, and also taking the investigation of the nature of imagination into the centre of theology itself.

MacDonald begins by offering us a definition of imagination. 'The word itself means an imagining or a making of likenesses. The imagination is that faculty which gives form to thought — not necessarily uttered form, but form capable of being uttered in shape or in sound, or in any mode upon which the senses can lay hold' (p.2). We might be tempted to speak of this faculty as *creative*, but we are reminded by MacDonald that it is so in no primary sense. In fact, 'a man is rather being thought than thinking, when a new thought arises in his mind. He knew it not until he found it there. Moreover, the forms which man uses in creative imagination are not themselves created by him; they are already prepared for him (e.g., the paints and canvas used by the artist; or the words and concepts used by the poet). In Shelley's words in *Adonais* ('Life like a dome of many coloured glass, Stains the white radiance of eternity'), 'has Shelley created this figure, or only put together its parts according to the harmony of truths already embodied in each of the parts?' MacDonald claims that it is a new embodiment, created out of forms already existing. The novelty is there, just as it is in God's creativity, insofar as it gives form to thought. In fact we should expect to find creative imagination in every sphere of human activity; notably, for example in the evolution of language, for 'the half of our language is the work of imagination' (p.7).

The key, however, to the kind of interpretation which MacDonald is going to give to the concept of imagination, and particularly to the emphasis which we have already observed he makes upon the element of creativity, is to be found in his very early identification of imagination as that feature of man's being which is the clue to understanding the nature of the image of God. 'The imagination of man is made in the image of the imagination of God' (p.3). In this way the name is given to this faculty in man on the basis of the nature of the power by which he was fashioned. MacDonald recognises that we can not say what form the thought of God takes which issues in creation; but he claims, however far-off, a resemblance, a clue, in the way our consciousness works.

Two points immediately interest me in what MacDonald says, and they will both be given the fuller development that they merit. First, he comes out boldly and claims what scarcely anyone before him, and very few since, ever had the courage or maybe, simply the wisdom, to declare namely, that imagination is an attribute, fully worthy, because of its involvement in God's creative activity, of its place alongside the other attributes to be found in any orthodox list of God's attributes. Secondly, he locates the *imago Dei* concept in an exceptionally novel place, not in man's rationality, or his moral character, or even his sociability, but in his imagination, and firmly links it to God's own creative activity. Even at a first glance we can see how the *imago Dei* so described pervades all of these other elements which appear in so many accounts of the *imago*. To continue with the exposition: At an early point in the essay (p.2) MacDonald had said that it is

INTRODUCTION 15

the function of the imagination to enquire into what God has made, as well as following out the divine function of putting thought into form. Now normally it would be claimed that the investigation into nature and its workings is an activity preempted entirely by science, a discipline in which it might be thought there is very little place at all for imagination. So, it might be concluded, imagination is excluded from any serious investigation of one whole vast section of God's imaginative activity, his created universe. But MacDonald will not have it, and replies with a number of arguments. For example, accepting that there is perhaps no place for imagination in the already ascertained field of accepted knowledge, he claims for it a role in relation to the undiscovered, the unexplored (p.11). Granted that the facts of nature are to be discovered only by observation and experiment, for which imagination can admittedly be no substitute, it is on the contrary the imagination which reveals which experiments might be done. Further, as every experiment has its origin in hypothesis, 'the construction of any hypothesis is the work of imagination' (p.13). The imagination can often catch a glimpse of a law before the detail of experimentation has had the opportunity to verify it. In fact, MacDonald goes farther and says that this light which the imagination is able to throw upon scientific problems derives from the circumstance of 'its inward oneness with the laws of the universe' — a remarkable suggestion both that the imagination operates under the constraints of laws, and that its own inner structure conforms to a rationality which matches the rationality of God and of his universe. The imagination is then all the time seeking to extend the boundaries of the pure intellect, searching out new territories into which to push what is perhaps a little unkindly called her 'plodding brother' (p.14).

It is, however, in history that MacDonald locates one of the most significant scenes for what he calls 'the intellectuo-constructive imagination' (p.16). 'To discover its laws, the cycles in which the events return, with the reasons for their return, recognising them notwithstanding metamorphosis; to perceive the vital motions of this spiritual body of mankind; to learn from its facts the rule of God; to construct from a succession of broken indications a whole, accordant with human nature; to approach a scheme of the forces at work, the passions overwhelming or upheaving, the aspirations securely uprising, the selfishness debasing and crumbling, with the vital inter-working of the whole; to illuminate all from the analogy with individual life and from the predominant phases of individual character which are taken as the mind of the people — this is the province of the imagination. Without her influence no process of recording events can develop into history'. This statement is quite remarkable, anticipating as it does, by some seventy years, the kind of thing that R. G. Collingwood was going to say on 'the idea of history' in the book of that name. Collingwood was to speak of 'interpolation', while MacDonald notes the construction, from a series of broken indications, of a whole accordant with human nature, that is with the motives that govern human behaviour and in particular with the character-patterns of the agent concerned. Such an emphasis upon interpretation of historical process on the basis of an understanding of human nature is in itself a corrective to analyses of history which concentrate upon movements and trends, forgetting that history is the action of human beings in their inter-relatedness and in their fragmented relationships. Pursuing this same theme, MacDonald makes an interesting but much too brief reference to the way in which these same principles of historical interpretation apply to the understanding of a human life, constructing the shape of that life out of pieces of documentation or oral tradition left from the past. 'How this will apply to the reading of the Gospel story we leave to the earnest thought of our readers' (p.18). Though this idea of MacDonald's might be thought to lead only to the development of the 'Jesus of history' approach to the Christological material, nevertheless it is a continuing reminder that to make what look like any historical statements about Jesus or even the New Testament records, is to carry on something of that imaginative, intellectuo-constructive activity which MacDonald describes.

Despite his interest in the theological associations of the imagination, MacDonald nevertheless does not neglect the traditionally acknowledged poetic activity of the imagination. It is not our purpose here to pursue MacDonald's themes into

INTRODUCTION

the field of aesthetics, for that is a subject worthy of independent research; and we have already mentioned the way in which the human imagination employs existing materials to present new thought-forms, which MacDonald says man does not create — they are given to him. But even this situation he presents in unusual theological terms, thus. 'God sits in the chamber of our being in which the candle of our consciousness goes out into the darkness, and sends forth from thence wonderful gifts into the light of that understanding which is his candle' (p.25). 'Thence we hope for endless forms of beauty informed in truth' (*ibid.*). Another ground for the connection of a work of art with this divine source MacDonald detects in the fact sometimes observed, that the work of art contains more than the artist himself intended, as, for example, in the variety of interpretations given to the same composition by different conductors, or to the same picture by different art critics. This affirmation of the divine inspiration of artistic activity has only fitfully been accepted in the history of aesthetics to the subsequent impoverishment not only of religious art, drama and music, but also of the liturgy itself.

Returning now to the religious interest: another noteworthy feature of MacDonald's treatment of imagination is the kinship which he draws between it and faith. Life has within it large spaces of uncertainty, and in them 'a wise imagination, which is the presence of the Spirit of God, is the best guide that man or woman can have' (p.28). Imagination, he had earlier claimed, has a particular capacity to deal with the fringe of the unknown and the uncertain which surrounds life. Quoting the biblical text, 'We walk by faith and not by sight' (2 Cor 5.7), he is virtually saying that imagination is the form which faith takes in face of the unknown. In exploring this hinterland, man learns to 'imagine greatly' like God who made him, himself discovering the mysteries, by virtue of an imagination which follows and worships God (p.29).

Because of the clearly religious and theological significance of the imagination for MacDonald, it is not surprising that he should seek to integrate it with the moral and devotional life. 'The first and essential means of the culture (sc. of the imagination) must be an ordering of our life towards harmony with its ideal in the mind of God. The man who is growing into harmony with God's will is growing into harmony with himself, all the hidden glories of his being are coming out into the light of humble consciousness; so that at the last, he shall be a pure microcosm, faithfully reflecting after his manner, the mighty macrocosm' (p.36). These ideas become the basis for advice on pedagogy, in which the role of imagination plays a central part — in teaching art-appreciation, history and, he might have added, science. But especially will the good teacher show that a great part of the Bible will only be rightly understood through the application to it of imagination as interpreted by MacDonald in terms of the *imago Dei*, the light lit within us by God himself through his Spirit.

Let me conclude with three fairly brief comments on the way in which MacDonald has advanced our discussion. First, he has lifted the plane of the discussion into a wholly theological dimension. Only in Coleridge is there any comparable recognition in the nineteenth century of the relationship between God and the imagination, and it was not until the second half of the twentieth century that the relationship was fully explored. Secondly, in tying the discussion in to a theological reference, MacDonald has not tied it to some esoteric fringe of the subject. He has linked it to the great central doctrines of creation, the Bible, the Spirit of God, the story of Jesus, the *imago Dei*, human goodness and personal piety, and the training of the young. Thirdly, MacDonald shows that he is willing to draw upon the philosophical and cultural components implicit in the notion of imagination, not as has so often been the case, to distort the faith, but to enable its fuller self-expression. The moral here is that we should show a like willingness to explore the potential of the concept as it comes to us out of its own very varied history, and so use it to explore, as MacDonald himself said, the dark places.

Chapter 2

THE PARABOLIC IMAGINATION

Before we proceed to examine the place of imagination in theology and to show how the philosophers may help, it will prove instructive to begin with the Bible itself, and to explore the subject of imagination as a biblical category first of all. Perhaps nowhere in the whole Bible is the basic character of imagination as a category of biblical thought better illustrated than in the parables of Jesus, nowhere better evidenced the implicit rejection of aniconistic thinking than in the way Jesus presents some of his fundamental teaching in the form of vivid images, pictures and stories. Reading the New Testament in short passages, either devotionally day by day, as sermon fodder or in public worship, we are seldom in a position to appreciate the sheer volume which the parables represent of the message of Jesus. When I speak thus of the images and the pictures of the parables of Jesus — and I use the word 'parable' in a sufficiently comprehensive sense to include the metaphors, similitudes, symbols and the occasional allegories, which are part and parcel of the parabolic presentation — I am not thinking of a highly contrived image system or world such as that of a Tolkien or a Lewis, the Hobbits, the Elves, the Wizards, or yet the Harfoots, the Stoors and the Fallohides, or any of the inhabitants of the Narnia sagas. On the contrary, the imagery of the parables is the stuff of ordinary existence, the everyday things in the everyday life of contemporary Palestinian Jews: the sower, the patient husbandman, the mustard seed, the budding fig-tree, the watchman, the playing children, the leaven, the treasure, the pearl, the seine-net, the lost sheep, the unfaithful servant, the labourers in the vine-yard, the two sons, the burglar, the talents, the last judgment, the two debtors, the good samaritan, the importunate friend, the rich fool, places at the banquet, the prodigal son, the unjust steward, the rich man and Lazarus, the unjust judge, the pharisee and the publican. The detail and the context are drawn invariably from the Palestinian life of Jesus' own time, and not infrequently authenticity of origin is evidenced in the use of the definite article where we would use the indefinite article, which according to Joachim Jeremias (*The Parables of Jesus*, ET, SCM Press, London, 1970, p.10), is common in parables and pictorial narratives in which the Semite thinks pictorially and has an image in his mind of a concrete instance, even although the reference is a general one. Gunther Bornkamm (*Jesus of Nazareth*, ET Hodder and Stoughton, London, 1960, p.69) gives a similar account of what we might call the 'ordinariness' of the substance of which the parables are constructed: Jesus' parables 'make use of the familiar world, a comprehensible world, with all that goes on in the life of nature and of man, with all the manifest aspects of his experience, his acts and his sufferings. Every spring and autumn, the sower goes over his field, every year wheat and weeds grow together, daily the fishermen catch good and bad fish in their nets. Often, to be sure, the events are such as do not, thank God, happen every day: the breaking-in of thieves in the night; the faithlessness and craftiness of a steward ... (p.70) By preference, and with great art, Jesus' parables tell just such stories, which are not by any means a regular feature of daily life. Yet (even so) they always remain within the realm of what every man understands, what is a daily or at least a possible experience. It is just this way that things happen'.

What makes the parables a particularly appropriate starting-point for the examination of the place of imagination within the Christian faith and theology is that they enshrine for us a particularly reliable part of the tradition, in reading which we are assured that we are in contact with Jesus himself, 'human life as worked upon and shaped by the creative imagination' of Jesus (G. V. Jones, *The Art and Truth of the Parables*, SPCK, London, 1964, p.113). Problems there may be for New Testament scholars about the original meaning of some of the parables; problems, too, about the reinterpretation of the parables in a modern setting; but we can be sure that in understanding how they are structured and how they work, we are glimpsing something of how the mind of Jesus acts imaginatively. Imagination, in this context, is a method, described as 'paradigmatic, comparative, homiletical and controversial' (by Jones, *op. cit.*, p.85), even a technique, of thinking, teaching, preaching, communicating, even praying, which employs imagery which everyone can understand. A case could almost be made for affirming that in the parables, as they appear in the Gospels, we are confronted by an originality which if not attributable to Jesus, can scarcely be attributed to anyone we know in the subsequent history of the Church. Our task, then, is to discover how what I would call 'the parabolic imagination' is employed by Jesus to accomplish his ends in the several activities we have mentioned.

That form of imaginative expression which the New Testament calls 'parable' has been, as was suggested, the subject of long-standing controversy, both in the Church at large, and understandably, still more so in the field of New Testament scholarship (cf. J. Jeremias, *The Parables of Jesus*, ET, SCM Press, London, Third Edition, 1972; D. O. Via, *The Parables*, Fortress Press, Philadelphia, 1967; J. D. Crossan, *In Parables*, Harper and Row, New York, 1973; edited by D. Patte, *Semiology and Parables*, The Pickwick Press, Pittsburgh, 1976; M. A. Tolbert, *Perspectives on the Parables*, Fortress Press, Philadelphia, 1979) but, by taking account for example of the different views that have been held of the nature of parables, and of what could be called their internal structure, we shall advance our appreciation of the role that they play in enabling us to understand the faith and to express it. For most of the history of the Church until recent times, the parables have been the subject of allegorical interpretation, which led to systematic misrepresentation and even distortion. Admittedly, there is evidence that this process of allegorisation had already begun within the Gospels themselves; and no doubt, it was thought that what was a valid approach to some, justified by the evidence mentioned, could without falsification be extended to all the parables. When that took place, the result was that the nature, purpose and message of the parables was obfuscated. But a break with a long-standing tradition of allegorical interpretation of the parables was to come with the work of Jülicher, in particular *Die Gleichnisreden Jesu* (J. C. B. Mohr, Freiburg, Leipzig, and Tübingen, 1899). Jülicher insisted that the parables be seen, as we have been indicating above, as 'a piece of real life' (Jeremias, *op. cit.*, p.16), but unfortunately he took off in another wrong direction. He attempted to elicit from each of the parables one single idea of the greatest generality. For example, Lk 16.19-31, with its picture of the contrasting states of the rich man and of the poverty-stricken Lazarus, was designed to bring happiness to those whose lot is to live in pain in this world, and to reduce to abject terror, all who here enjoy and indulge a life of ease. The parable of the rich fool (Lk 12.16ff) affirms the generality that even the richest of men is at every moment entirely dependent upon the power and mercy of God. Jeremias offers two objections to Jülicher's handling of the parables; first, that he strips them of all their eschatological significance, giving them a wholly human content; and, secondly, he presents Jesus as a teacher of wisdom, who dispenses moral sayings and reflections by means of some very appropriate and telling metaphors and stories.

Two comments would here be in place, though they are not offered in justification of Jülicher, who is vulnerable at the points indicated by Jeremias. First, there are scattered throughout the Synoptic records of the parables, a not insignificant number of generalised sayings, as Jeremias acknowledges, for he devotes a lot of his time to showing how they have come to be transposed, and how they have been differently employed by the different Synoptists. Jülicher is not therefore totally out of order in recognising the presence of these generalised sayings. Secondly, Jülicher draws our attention to a matter of considerable importance, namely, the universalising tendency which is combined in the parable with the particularity of each parable. In fact, as our discussion advances, we shall begin to observe that one of the hall-marks of all imaginative thinking, and not just of the parabolic imagination, is the way in which through the specificity of images, similes and stories, it achieves universality of implication and of application. At the moment, we simply note the point, and give notice.

But Jülicher's chief error lay in his assumption that the parables could be cut loose from their moorings in the precise detail of the situation to which they were directed, and allowed to drift free and attach themselves to any moral occasion that subsequently arose in the history of the church. On the contrary, not only did they occur in actual situations in the life of Jesus, they had something to say to these precise situations and to the people in them — often some word of correction, reproof, challenge, attack, or forgiveness (Jeremias, *op. cit.*, p.19). So, says Jeremias (*ibid.*) 'Each of (Jesus') parables has a definite historical setting ... Jesus spoke to men of flesh and blood; he addressed himself to the situation of the moment'. To continue a little farther with Jeremias' reflections on the parables we note his rejection of the endeavours of the Form-Critics to classify the parables according to categories. Such labour, he says, is fruitless, for the Hebrew *masal* and the Aramaic *mathla* embrace all the classes of parable, similitude, allegory, fable, proverb, oracular utterance, riddle, significant name, symbol, pseudonym, example (type), commonplace, argument, apology, refutation, jest. So, the New Testament *parabole* means parable or comparison (Lk 5.36, Mk 3.23), symbol (Heb 10.9, 11.19, Mk 12.28), proverb or commonplace (Lk 4.23), riddle (Mk 7.17) and rule (Lk 14.7). It does therefore appear hard to understand what purpose is served by the Form-Critical analysis of the types of parables, especially when it is derived from Greek rhetoric and applied to rather alien subject-matter. Perhaps, also, it is not unfair to note, the ineffectiveness of the classification mentioned is due to the fact that such categoreal descriptions (which in normal Form-Critical analyses are calculated to expose some purpose served by a redactor or even the Primitive Church in imposing some particular form) add very little in any case to the content of the parable, whereas the purpose of the parable is to be found not in some externally imposed scheme or some extraneous intention of a writer or editor, but in the situation to which it was initially directed. Attention has been drawn to this view of Jeremias', because it would constitute evidence in support of the premise from which we took our start earlier, namely, that we were not proposing to draw unduly sharp distinctions between parables proper, and the images, metaphors and similes which are often comprised within them, or appear independently in the biblical text. Another point which emerges from the study of the history of scholarship on the parables, and is of interest to us here, is that, admittedly the rabbis had employed parables extensively in their teaching and so the parables might not — contrary to what we said earlier — be original to Jesus' teaching. But in the method of the rabbis the parables were intended to clarify some point in their teaching, to help in the interpretation of a passage, and generally to be of assistance in the exegesis of an authoritatively prescribed text; whereas, while some of Jesus' parables resemble in content those of the rabbis, in his use and as he deploys them, the parables are in many cases themselves the message, or as Bornkamm (*op. cit.*, p.69) calls it, 'the preaching'. They are not allegories or illustrations of some theme, which is stated somewhere else and non-parabolically. What they are about is to be found in their content and in their context. Take the case of the many parables about the Kingdom, or the Reign of God: these parables themselves constitute what Jesus has to say on this central theme of the Gospel and the Gospels. We must not regard them as offering us an optional extra, otiosely additional to some non-parabolic aniconastic account which appears somewhere else in the scriptures. The message concerning the Kingdom *is* the parables about the Kingdom. Bornkamm makes this same point with even more force and intensity when he claims (*op. cit.*, p.25) that each of the story scenes or anecdotes in which the Gospels tell the story of Jesus 'contains the person and history of Jesus in their entirety'. No elaborate explanation requires to be drawn from previous occasions or passages, nor have we to await some further events before the meaning of an earlier event can be unfolded. Each scene is complete in itself, as the circle of light illumines and defines the event and the persons in the story. That fact is equally true of the words of Jesus, each exhaustive in itself, and not demanding commentary or elaboration from some other word or saying.

Let us now proceed to examine the structure of the parables as we have them in the Gospels, for they have light to shed on the nature of imaginative thought and communication. At the outset we are warned that the scriptural introduction to the parable may in fact mislead us as regards the other entity with which the subject of the parable is being compared. This circumstance is due to the continuance into the Greek New Testament of a form, namely, the Aramaic dative *le*, used to introduce rabbinic parables. The fully extended form would read: 'I will relate a parable to you. With what shall the matter be compared? It is the case with it as with ...' So we have in Mk 4.30f: 'With what can we compare the Kingdom of God, or what parable shall we use for it?' The answer follows: 'It is like ...' which Jeremias, whom we are here following, says should be translated as 'It is the case with it ... as with ...' When we apply this formula to actual parables, we see that we have to exercise caution in our naming of the entities actually being compared in the parable. For example, in Mt 13.45, the Kingdom is not like a merchant, but a pearl; in Mt 25.1, not like ten virgins, but a wedding; in 22.2, not like a king but a marriage feast; in 13.24, not like a man who has sowed good seed, but like the harvest; in 20.1, not like a householder, but like the distribution of wages. In fact, I think that I would like to take the matter farther than Jeremias here, and say that the comparison is between the Kingdom and, not any single entity, but the whole situation deployed in the anecdote. The different items mentioned by Jeremias in relation to these several parables could be regarded certainly as the nucleus of the anecdote, but they require the amplification supplied by the rest of the anecdote to give them meaning, and to supply the detail to support such significance. In fact, Jeremias seems to concede this point when (at p.79) he says of Mt 13.47: 'the Kingdom of Heaven is not compared to a seine-net, but the Kingdom at its coming is compared to the sorting out of the fish caught in the seine-net'. One caveat ought to be entered at this early point of our discussion: it is directed against assuming too readily that when we think of the situations referred to in the parables, we necessarily employ detailed pictures of these situations. The word 'picture' occurs from time to time in the literature on our subject, but there is evidence to suggest that there is not full pseudo-photographic recall or construction involved in this type of thought. The images seem to carry a form of representative tab, which allows us to think of the whole by means of the part — what I would call 'synecdochal thinking'.

One particular problem does arise from a certain aspect of the parables which we have stressed, namely, that they are specific to the situations of their origin; their message, in the first place, has immediate relevance to these situations, and they are not to be read as universal maxims. Now it is plainly the case that not a few of the parables are presented together with just such universal statements. Lk 18.14b: 'Everyone who exalts himself shall be humbled, but he who humbles himself shall be exalted'. Lk 13.30: 'Some are last who shall be first, and some are first who shall be last'. Lk 16.13 'You can not serve God and mammon' (cf. Lk 19.26; Mt 25.29). There are several ways in which we handle this problem, as problem it is if we are going to hold to the view that the parables are not simply illustrations of some generalisations, or expendable ornamentation to a quasi-Aesopic moral. First, we may tackle it simply as a critical problem, and look at the detail of the text in each occurrence, attempting, by comparison with parallel passages in other Gospels, or even within the same Gospel but elsewhere, or by consideration of possible hortatory or allegorising motives, to establish the primitive form of the text. Such examination will on occasion yield the result that the generalisations which are appended to the anecdotes that form the parables, were themselves genuine *logia*, but not originally in their present position, being put there to effect the application of the parable (cf. Jeremias, *op. cit.*, pp.85f). There are two reactions to this position which is associated with the name of Jeremias, and I think that they both tell us something about how the imagination works. The first reaction accepts the idea that the generalising *logia* are not part of the parables, but are rather addenda which spring from a failure to understand how the parable 'works', especially when taken along with other parables on a similar subject. The practice of adding a generalising *logion* arises from the feeling that the parable, in its original form, has to be interpreted, or explained, or applied; and that the *logion* will act like the draw-cord of some cautionary tale. It may even represent the aniconastic tendency which we have observed at work in reflective theology, the conviction that truth does not reside in the image or the anecdote, but only in the concept and the abstract truth, never in the specific, but in and only in, the universal. If, therefore, we can succeed in subtracting the universalising *logia*, and look at the parables, about the Kingdom, then the technique of imaginative expression in leading us to the truth, is to heap image upon image, in rapid profusion, in the hope that, as Ian Ramsey so often said, the penny will drop, or the bell will ring, and we say 'I see it all now'. That seeing is the end-term of the process, and it is not to be codified into some abstraction. We run from particular image to particular image and so to the conclusion that Jesus wants us to see — but the conclusion also is particular, and as we shall see later, it is directed particularly at us and not at the generality of mankind. In fact, the possibility of the parable *not* being directed to us is ruled out from the very start in that not a few of the parables begin with the words, 'Which of you ...?' — a form absent from rabbinic lore, according to Bornkamm (*op. cit.*, p.70) who adds 'It is always a question aimed straight at the hearer himself, which neither demands from him knowledge or theoretical judgment, nor presupposes his goodness or education'.

But there is another point-of-view on this subject of particularity and generality in the parables, and it is expressed by Rudolf Bultmann (in *History and Hermeneutic*, by Wolfgang Pannenberg *et al.*, Bultmann's essay being entitled 'General Truths and Christian Proclamation'). Referring to a number of the cases we have already mentioned, Bultmann asks, 'And is not the *tertium comparationis* of many of (Jesus') similitudes and parables a general truth, however often the modern interpretation of the parables has refused to recognise this?' Bultmann has a very special interest in this subject in the paper referred to, for he is anxious to establish the case against the view that the parables did not have the character of proclamation insofar as they appeared to expound general truths (*op. cit.*, p.153). 'Proclamation is personal address'. Something is happening, there is genuine communication, in a sermon, the event being the revelation of God's grace which has occurred in Jesus Christ, and which is now occurring in the proclamation, that is, the sermon, where the revelation of Jesus Christ takes place anew. If, then, the Christian sermon is a message which 'is spoken as an address into the immediate moment', how do we reconcile this account with the fact that there are in the proclamation of Jesus what appear to be many general truths? So, too, in St Paul there are many statements which have a proverbial character (e.g. Gal 6.7; 1 Cor 5.19).

Bultmann's resolution of the problem is interesting, and it is as follows. Though these general truths are such as a man could say to himself, yet they do share in the address-character of the proclamation insofar as the proclamation, by means of them, addresses to the hearer some truth or truths that he may well have forgotten, and does so to his own shame and condemnation, or to his comfort and encouragement. The purpose of this reminding is that through the proclamation, he may be brought to the point of decision, and 'thereby be transformed'. Bultmann would be prepared to include ethical imperatives in such an analysis, because, for all their generality, even they may appear within proclamation and so be addressed here and now to a particular hearer in a particular situation. Bultmann's interest is, therefore, predominantly kerygmatic, and his views can not be taken as an altogether non-doctrinaire, completely authoritative basis for the exegesis of the parables. Nevertheless, he has warned us against a total removal of the generalised *logia* from all connection with the parables, either in their original form or in a form adopted by the Primitive Church to give them a wider application, or a hortatory intention. Also, he brings out something to which we have probably not so far given sufficient attention, namely, the unique way in which the parables combine the direction to a specific situation and the specific communication to that situation *with* a universal significance. It is just here that I find Jeremias somewhat unsatisfactory. Admittedly, he does recognise that the parables have what he calls 'a double historical setting', on the one hand, of some specific situation in the pattern of the activity of Jesus, most of the parables arising out of some such actual occurrence; and, on the other hand, of the proclamation, preaching and teaching of the Primitive Church, which, before the words of Jesus were given a written form, collected and arranged them according to their subject-matter, and created a new setting for some of them, expanding, allegorising, and as we have seen, universalising others, by adding the *logia*. In this distinction, in which incidentally Jeremias is anticipating what was to be so

THE PARABOLIC IMAGINATION

heatedly debated under the rubric of 'plurality of meaning', Jeremias clearly seems to be assigning the universalising element in the parables to the action of the Church, whereas, in my judgment, that element was present and implicit in the original situation and message of Jesus. I would offer two reasons for this statement: first, as we shall now be observing, Jeremias argues at considerable length and with appropriate documentation that the major part of the teaching of Jesus has come to us in and through the parables, which in itself illustrates the trans-specific universally significant quality of the parables; and secondly, if we are right in regarding the parabolic creations of Jesus as the expression of the religious imagination, we should expect them to contain both a particular and a universal import. In fact, it is of the essence of a parable to be universal, to be transitive, simply because of its particularity. The paradigm case is the parable told to a certain lawyer (Lk 10.25-37), who, enquiring what he should do to inherit eternal life, received in answer the story of a man who went down from Jerusalem to Jericho and fell among thieves, being rescued by a Samaritan, and was given the command, 'Go, and do thou likewise'. The lawyer was expected to make the transition from the story, in all its non-allegorical particularity, to his own private and personal circumstances; and to 'see the point' of the story in what he, and no one else, was required by Jesus to do.

Let me return to the case I was making above for the theory that the fact that the major part of Jesus' teaching was given in parables is in itself evidence of the universal element being implicit in the parables, present there without the need of a special indicator such as a *logion* to advertise the fact. Jeremias had said, 'Jesus never tired of expressing the central ideas of his message in constantly changing images' (*op. cit.*, p.89). Just how comprehensive this presentation was we may readily see if we look briefly ('briefly', because Jeremias devotes half his book to this part of his exposition of the parables, pp.115-229) at some of the ten categories into which he groups the components of Jesus message. I have selected the following: The message that 'now is the day of salvation', what a later theology was to call 'realised eschatology', is presented in a whole variety of parables — the shepherd going to the lost sheep and bearing it home, and the shepherd giving his life for his flock; the physician who has come to heal the sick; the fisherman appointing fishers of men in his service; and the wedding parables carrying the eastern symbolism of the Day of Salvation. The Good News of God's mercy for sinners is contained in the parables of the lost coin, the lost sheep and the lost son, the latter particularly describing 'with touching simplicity what God is like, his goodness, his grace, his boundless mercy, his abounding love' (Jeremias, *op. cit.*, p.131). But Jesus does not only proclaim the day of Salvation; he announces also judgment and warning, and summons to repentance, this time through the parables of the children in the market-place, the unfruitful fig-tree, the rich fool who prepares for an even heavier harvest, the servants entrusted with authority when their master went into a far country, the talents and how they were variously used. The demands of discipleship are stated by Jesus in terms of total commitment, which he expresses in the parable of the pearl of great price. It entails boundless love towards others, expressed both in the parable of the Good Samaritan, and of the sentences pronounced at the Last Judgment (Mt 25.31-46). So discipleship is marked by three characteristics that have been also described in parabolic terms. First, it will extend to others the forgiveness which it has itself received (as witnessed to in the parable of the unmerciful servant in Mt 18.23-25). Secondly, the disciples will have the assurance of absolute security in God's hands, like the birds of the heavens in Mt 6.26, or the flowers of the field, as in Mt 6.26. Thirdly, the gift of God and the call of Jesus impel to action; and despite the enormity of the task, Jesus will not allow them to be dismayed by it. If they have faith even as a grain of mustard seed, they will have the power to move mountains. In the same way, when he comes to speak of the *via dolorosa* which the Son of man must tread, it is again to metaphors and parables that Jesus turns, describing his destiny as the cup that he must drink, the baptism with which he must be baptised. He is the corner-stone that must be rejected, the corn of wheat that must die. His death is to be a ransom for many, a sacrifice whose blood is shed for the redemption of the many. A veritable avalanche of imagery and of parables is employed to present the final consummation

THE PARABOLIC IMAGINATION

of the purposes of God for his people and for the world. God the King will be worshipped in a New Temple (Mk 14.58), with the Son of Man sitting at his right hand surrounded by angels. Evil has disappeared and the sinful world is no more. Satan has been cast down from heaven. Death is no more, suffering and sorrow have come to an end. Sinners are forgiven, but the good and bad must remain to the end, growing together until the harvest, and being allowed to ripen in patience.

The greatest single point that comes home to us in this review (inspired by Jeremias) is the way in which, you might say, the whole of the essential teaching of Jesus is both contained within the parables, and mediated to us by powerful and moving imagery and imaged story. Through a whole range of anecdotes and images, whose components were the ordinary stuff of the day-to-day existence of his contemporaries, Jesus conveyed to his hearers all, literally all, that he had to tell them about the nature of God and his love and care, about the will of God for them, about the purpose of his coming into the world, about the kind of death that he would die and the central purpose of it, about nature and history, and about the end of both the world and history. It was his intention, too, that his hearers should grasp what he had to say in the terms in which he said it, and not require the assistance of the vast hermeneutic machinery which two thousand years of tradition scholarship have invented, allegedly for the elucidation of the text, but often, in fact, for its concealment, or worse, its distortion. Now it would admittedly be somewhat anachronistic to try to turn the wheel back to the reproduction in our time of that kind of theological expression and exclusively to that style of presentation. Also, to try to do so would perhaps not result in theology as we know and practise it. Nevertheless, you are still left with the very uneasy feeling both that we have not done justice to the possibilities of iconastic thought with the thoroughness which the example of Jesus demands that we should, nor have we been sufficiently rigorous in our pruning of the theologoumena with which we have allowed theology to become cluttered in this as in earlier centuries. Iconoclasm such as has been experienced in our Scottish churches has been matched and even outstripped by de-iconising our religious expression to the consequent impoverishment and even barrenness of our teaching and preaching and writing. I can not escape the conclusion that through the vast volume of parable, metaphor, simile and image, Jesus is saying something to us about how we should be talking and thinking about the fundamental facts of the faith; and that we have refused to listen, or listening, have been unable to hear because of the conditioning of centuries of other voices telling us of other, more complex, ways of carrying out these activities.

Discussion of the relation of imagination to the parables would, however, be incomplete without some account being taken of those actions of Jesus which are known as 'acted parables'. Bornkamm (*op. cit.*, pp.157f.) has provided a convenient listing. Jesus' choice of twelve was symbolic of his sovreignty over the eschatological people of Israel; his entry into Jerusalem, and his cleansing of the Temple gave symbolic expression to his kingly authority; in choosing an ass to convey him into the city, he symbolised the peaceful purpose of his mission; he washed his disciples' feet as an example of how they should honour one another; the writing on the sand was a parabolic action to remind the woman's accusers of the duty of repentance; while the taking of the common meal on the night before his death afforded Jesus the opportunity to 'perform the last symbolic act of his life', offering thereby to his disciples 'a share in the atoning efficacy of the death that awaited him'. That is the normal interpretation of the acted parable. Jesus does a certain action which has two meanings: the *prima facie* and superficial meaning — riding on an ass, washing feet because your friends have just completed a very tiring journey, and so on; and the deeper meaning, the transferred reference to some other reality — the Reign of God, the atoning death, the duty of caring — all meanings out beyond the immediacy of the actions themselves. We have, then, on the one hand, the visible action; and on the other, the parabolic significance of it which is not actually stated in the first instance, at least, though it does appear later as a kind of explanation for someone who has failed to see the point.

It is that 'normal interpretation' of the acted parable which I wish now to question, on the ground that it dichotomises

THE PARABOLIC IMAGINATION

what is in fact a unity. There are not two things, the action and the parable, the one separated from the other by the relation which connects symbol and symbolised, example and the point being illustrated by the example. The acted parable is a unity, and it is so in virtue of what I call the 'realistic imagination', which sees *as one* what we divide into the two components in a situation with an external reference. Realistic imagination appears at many points in the biblical narratives, though in our reading and interpreting them, we constantly try to substitute for it a symbolic hermeneutic. Take the well-known story of Elijah and the four-hundred and fifty prophets at Mount Carmel (1 K 18.20-40). We would normally regard the final denouement of the story to be that the consuming of Elijah's offering by fire was an empirical demonstration to all the people of Israel there gathered, of the truth that Jahweh was the one and only true God — a sign symbolising the universally valid proposition, an illustration of a fact that was not part of the event. On the contrary, for the people there gathered as for the writer of the story, the supremacy and Lordship of Jahweh was realistically part of the event. They had the kind of imagination which saw the Lordship in the happening, and not residing, as it were, at some point external to it — in the mind of the narrator, or the thought of the beholder or the reader. In much the same way, the anointing of Jehu in the story in 2 Kg 9.1-13, by the young prophet sent by Elisha, was not simply a symbolic act which would have to be ratified once the kingship had been verified by the processes of constitutional investigation. It was an acted parable, understood by the realistic imagination as effecting its end, so that when Jehu rejoined his colleagues, one and all they spread their garments upon the ground and blew the trumpet, and proclaimed, 'Jehu is king'. In order to arrive at the distinction between the symbol and the symbolised, it is necessary to introduce a division in what is essentially a unity for the realistic imagination: anointing Jehu and making him king were one and the same thing. Something of this same unity is to be seen in the story of Jacob's obtaining Isaac's blessing with the trick of dressing up in skins and preparing a fine meal for his father, pretending to be Esau who would have been the proper recipient of the blessing. We would be rather inclined to separate the words of Isaac in which the blessing was couched and through which it was to be conveyed, from the blessing proper. We would introduce such notions as 'mistaken identity' and 'false impersonation with intent to deceive', and argue that therefore, though the words were spoken, the reality which those words would normally convey was invalidated by the moral dishonesty and deceit. But for Isaac, as for Jacob and Esau, as well as for Rebekah, the instigator of the whole plot, the speaking of the words by Isaac was *eo ipso* the actual blessing, and it could not be cancelled or recalled as might some form of words only. The words and the blessing were a unity, not an outward sign and an inner spiritual reality.

It is just such a separation which, as I argued some years ago (*On the Love of God*, Collins, London, 1962, pp. 109f.), has caused difficulties over the years in the definition of the theology of the sacraments, particularly of the Lord's Supper. Western thinking about the sign and the signified, symbol and the reality which it designates, has introduced division where there originally was unity, and so created the difficult problem of saying how these two entities thus separated are related to one another. The very variety of the ways in which theologians have described the relation should have made us ask prior questions about their analysis of the situation which induced such sophisticated interpretations of the relationship. One group said that the signs serve to call to remembrance what they signify, the bread, the body of Our Lord and the wine, his blood; another, that the relation is one of external compresence, so that receiving the elements, we receive 'in, with and under' them, the body and blood of Jesus Christ; yet another, that through the faith given by God himself, the communicants in receiving the elements receive with them the body and blood of their Saviour; while, finally, another theology will wish to say that bread and wine become the food of salvation through an exchange of substance ('transubstantiation') with the body and blood of Christ.

Now we normally take our stand on one or other according to our denominational loyalty; but the question which I would like to ask is whether we might suspend such loyalty to see if there is some other way of looking at the situation, one which

THE PARABOLIC IMAGINATION   35

did not commit us to stating relationships between divided entities, which should not perhaps have been so rigidly separated in the first instance.

In order to present this new look, I want to go back to a circumstance that we observed in the Old Testament stories of Isaac, Jacob and Esau; of Elijah and the prophets of Baal; and of Jehu and his anointing; and follow it through to what I regard as the paradigm-case for the understanding of the Lord's Supper. I refer to the account of David at the cave of Adullam (1 Chr 11.15-19; see also *On the Love of God, loc. cit., supra*). David was at that time in control of the stronghold, the city of Jerusalem, but the Philistines were engarrisoned in Bethlehem. With longing David said, 'O that someone would give me water to drink from the well of Bethlehem which is by the gate' (v.17). Three of his warriors succeeded in penetrating the lines of the Philistines, in drawing water from the well and returning through the lines with it to David. When they gave it to him to drink, David refused and poured it out on the ground 'to the Lord' saying, 'Far be it from me before God that I should do this! Shall I drink of the life-blood of these men? For at the risk of their lives they brought it!' (v.19). At the moment of pouring out the water on the ground, David was not unaware that had he drunk the water, it would have tasted exactly as he expected, remembering its coolness and freshness from his boyhood years, and would not have tasted like blood. Nor on the other hand, however, did he feel prompted to say that while remaining clear, pure and fresh, it symbolised the sacrificial spirit of the warriors who took their lives in their hands when they went out to draw it. Once again, the realistic imagination of David made an immediate identification of water and blood, so that in pouring forth the water, he could say, 'Is not this the blood of the warriors?' No question arose of how two entities, being separated from one another at the start, could come subsequently to be related. The whole statement about the blood of the warriors takes its departure from the identification in the imagination, of water and blood.

Admittedly, when we come to understand the words of Jesus at the Last Supper, we may be in error in seeking to do so on the basis of a number of Old Testament incidents, and of one saying, however similar, in the Old Testament. The blood of Jesus has very specific salvific denotations in the New Testament, which could never attach to the blood of the warriors. But it is not similarity of content of the two sets of statements that we are affirming, but of use. We are saying that when Jesus spoke of the wine and his blood, he was employing a usage which we have already encountered in our Bible-reading. I feel, too, that there is a great deal more sense as well as much sounder theology, in looking in that direction, than to the sophistications of western philosophy. What then happens when we apply the paradigm-case of David and the blood of the warriors to the words of Jesus? Immediately we can say that for Jesus, as for David, had he drunk the liquid, there would have been no question about it not tasting like wine, or not having the normal qualities of wine. Nor, on the other hand, did he, seeing the wine, resolve that it should henceforth 'stand for' blood, or symbolise it in some specially appropriate way. Nor, to add a further interpretation which might have come from D. D. Evans (*The Logic of Self-involvement*, SCM Press, London, 1963), or even from Wittgenstein, did Jesus look on the wine as blood, or see it as blood. Even that language, which is an improvement upon such dualisms as sign/signified or symbol/symbolised, starts from an assumption of separation, depends for its acceptance upon the continuance of the separation throughout the process of 'looking-on' or 'seeing-as', but never achieves the unity which it had destroyed from the start. When Jesus said, 'This is my body which is broken for you; this cup is the new testament in my blood', there was an immediacy of identity effected through the realistic imagination, of bread and body, of wine and blood. In receiving bread and wine, the disciples have been guided by Jesus' own use of realistic imagination, and in receiving, they too accept the identification. In so doing, they equally realistically enter into the new covenant achieved through the atoning death of Jesus Christ. The phrase 'immediacy of identity' has been used and it may repay consideration. 'Immediacy' here carries the connotation of being unmediated, that is, by such relationships as 'symbolising', 'signifying', 'illustrating', or even 'revealing', though the last term comes closest to what we are trying to say — very much

THE PARABOLIC IMAGINATION

as the humanity of Christ is both an element in the Revealer and a component in the Revelatum. In conclusion, let me make the following comments upon this construction of the meaning of the words of Our Lord at the Last Supper.

First, given that this special sort of identity is the truth about the relation of bread and wine to body and blood of Christ, it is not difficult to see why subsequent theology, which has taken as its starting-point the separation of the bread and wine from the body and blood, has experienced so much embarrassment in achieving a satisfactory account of how they are related to each other. There is a sense in which the words 'What God hath joined, let not man put asunder' are particularly appropriate to the theology of the sacraments. In fact, if there is truth in what has been said, we might be tempted to add, 'A plague, not just on both your houses, but on all of them', insofar as you have set yourselves the impossible task of trying to relate two sets of entities which exist in an unity which your categories can not grasp.

Secondly, if we were to firm up what we have been saying so that it is not only an interpretation of what Jesus said at the Last Supper, but the beginnings of a theology of the sacrament; then one of the questions which we have to ask — and answer — is: Is it being proposed that realistic imagination should replace faith as the human response in the sacraments? The answer is an immediate and categorical 'no'. Faith does much that imagination can not be expected to do: for example, obey, trust, acknowledge, respond, decide, to mention but a few. What is being said is that a genuine contribution can be made to our understanding of the nature of faith, as well as of the subject to which faith is related in the different ways enumerated above, if we allow that imagination, realistic imagination of the kind that we have been describing, has a proper place within it.

Thirdly, it will still be realistic imagination which enables faith to see bread and wine in immediate identity with the body and blood of Christ, as was the case 'on that night in which he was betrayed', when he first broke bread, and blessed wine, in this very special way. That original situation is repeated as often as we fulfil his command to do this in remembrance of him. If the command is faithfully fulfilled, then the circumstances that confront us will not have altered with time; and on his part, he will keep his promise to us.

Fourthly, if we concentrate now on the words 'This do in remembrance of me', we may note the relevance to them of Sartre's idea that one of the functions of imagination is to picture the past-as-present, and when it does so realistically, it is in fact dealing with a present reality, a 'real presence'. This element in imagination we shall consider later for it has great importance for the way we read the Bible, for understanding the person and work of Christ, and for our devotions.

Since *anamnesis*, remembering, is not the whole of the sacrament, though a necessary condition of its happening, the imagination has the further role of picturing the absent-as-present in a very special way appropriate to the sacrament. The point is not an easy one to make, and it has to be taken carefully to avoid misunderstanding. There is an obvious sense in which Jesus Christ, risen, ascended and glorified, is constantly and everywhere present, and in that sense he can not ever be absent, nor can we ever expect him to be more present than he constantly, and in his own right, is. It is the matter which Robert Bruce faced in his famous sermons on the sacrament of the Lord's Supper, preached in St Giles', and which he resolved in this way, 'In the sacrament, I get a better grip on Christ nor I get elsewhere'. In our language, the sacrament is the means, or one of them, which Christ uses to make himself especially available to us; and on the human side, the imagination is the means which he employs to achieve that end. So, if we were to correct what we have already said about the imagination in this connection: it is the means employed by Christ himself who is ever present to us, to make himself still more really and more nearly present and more truly available.

Sixthly, when we add to what we have said, our previous emphasis upon synecdochal thinking in imagination, a new aspect of its functioning in the sacrament appears. In synecdochal thinking the imagination, on perception of the part, immediately 'runs' on to complete the whole. So in the sacrament of the Lord's Supper, when that part of the original unity of bread and wine, and body and blood which is visible, is presented, and the words of institution are heard, the

THE PARABOLIC IMAGINATION

imagination 'runs on' to complete the whole; and Christ once again is present to his faithful people.

Seventhly, a question might be asked why the imagination is singled out for this specific role in relation to the sacrament of the Lord's Supper. I am reminded of an answer which David Hume gave when a similar question was raised about imagination in relation to our belief in causality. He said that there were three possible grounds for that belief, namely, the senses, reason and imagination. He ruled out the senses, as we must do also. The senses perceive only bread and wine. He ruled out reason which can reach only as far as regular succession. Again, so must we, for there is no process of ratiocination, no deductive or inductive argumentation, no compulsive logic of an objective reality, which will allow us to reach no other conclusion from receiving bread and wine, than that they are the body and blood of Our Lord. It is realistic imagination which makes the immediate identification, following the example of Our Lord in instituting the rite.

CHAPTER 3

IMAGINATION AS A THEOLOGICAL CATEGORY

Having considered the quite considerable place which imagination occupies as a category of biblical thought, particularly in the parables of the New Testament, appearing so extensively in the teaching and actions of Jesus, we can not but be surprised to find that it is relatively absent from the field of theological subject matter — though, as we shall see, not from the field of theological method. It was noted earlier that aniconastic thinking had prevailed in the history of theology, through the widespread employment of conceptual thinking to replace image thinking — a fact all the more strange to find perpetuated into times when theology has taken such pains to demonstrate how 'biblical' it has been. What, therefore, we are setting out now to do, following the lead of George MacDonald, is to discover points within the traditional subject-matter of theology, where we may introduce this category, without producing any distortion or falsification of doctrine, but perhaps, rather, achieving fresh insights into familiar topics.

Imagination and the Attributes of God

We shall begin with a consideration of the possible place of imagination within, or in relation to, the attributes of God. To this end Karl Barth's discussion of what he prefers to call God's 'perfections' (*Church Dogmatics*, II/1, ET, T & T Clark, Edinburgh, pp.322-677) has been chosen to introduce the subject. The reason for this choice is that *a priori* we might be inclined to think that that would be a very difficult area in which to discover such connections. We would hope in this way to establish our case in what can not be regarded as a soft option. From the various traditional descriptions of God's attributes as *appellationes* (titles), *virtutes* (excellences), *proprietates* (properties), *attributa* (attributes), and *perfectiones* (perfections), Barth selects the last, on the ground that, whereas other entities and beings may have attributes and so appear to share this characteristic with God, God alone is identical with the perfection of his attributes. While in the end it proves entirely impossible to avoid altogether the use of the term 'attributes', Barth will never lose sight of his chief intention not to allow any line to be drawn between God and his attributes, such as does exist in the case of other entities and beings. 'There is no possibility of our knowing the perfect God without knowing his perfections' (*op. cit.*, p.322). Barth's account of the knowledge of such perfections is set within the truth, which we encounter in so many of his writings, that the one God is for us fully revealed and fully concealed in his self-disclosure. This unity and this distinction correspond to the unity and the distinction in God's own being between his love and his freedom. He is in himself the one who loves us, and so is completely knowable to us. But he loves us in his freedom, and so in his revelation he is known as sovereignly free; nevertheless in himself he remains completely unknowable. So for Barth there arises the necessity to treat the attributes of God in a twofold series. The two fundamental features of the being of God indicated above — his love and his freedom — indicate the twofold direction of the ensuing discussion: the perfections of the divine love and the perfections of the divine freedom. Barth's perception of the relationship of God's self-revelation in Jesus Christ to his self-concealment in that disclosure is one of his major contributions to the theology of revelation, and if we do not delay here to expand upon it, that is not due to any under-estimate of its importance. Equally we shall resist being drawn into an examination of Barth's correlation of God's love with his knowableness, and of God's freedom with his unknowableness; for these correlations do not jeopardise Barth's classification of the divine perfections, which read as follows: 'The divinity of the love of God consists and confirms itself in the fact that in himself and in all his works God is gracious, merciful and patient, and at the same time holy, righteous and wise' (*op. cit.*, p.351). The other classification is as follows: 'The divinity of the freedom of God consists and confirms itself in the fact in himself and in all his works God is one, constant and eternal, and therewith also omnipresent, omnipotent and glorious' (*id.*, p.440). The matter is not, however, as simple as at first appears, namely, that Barth is describing, first of all, the six attributes of the God who is loving; and next, the six attributes of the God who is free. He complicates the situation somewhat by adding that within the first series, wisdom in God is related to patience, as holiness is to grace, and righteousness is to mercy; and, moreover, that 'the second set of ideas — holiness, righteousness and wisdom, express more clearly than the first — grace, mercy and patience — that it is his distinctively free divine love of which we are speaking' (*id.*, p.422). Of course, no antithesis is intended between the two categories of attributes in the first series of six. The recognition of the depth and majesty of the divine grace, mercy and patience calls for a special consideration of his holiness, righteousness and wisdom. When he deals with the second series of six attributes, a similar correlation of the two sets of three attributes emerges, and recognition of the attributes of the divine freedom in their depth and majesty will entail special consideration of those attributes which characterise and describe the freedom of God as the freedom of divine love.

Suspicion may well have been roused by this time that in such a complex theory of the divine attributes, one also which is so close-knit, meticulously patterned and integrated, it will be extremely difficult to find a place for so undogmatic and unusual a concept as imagination. Certainly, if we are looking for the word we shall not find it. Nevertheless, there are points at which Barth's account of the attributes comes very close to implying the presence of the concept of imagination, and where no distortion, or diminution, or caricature of meaning will occur if we do apply the concept. Moreover, by discovering the concept in these contexts, we shall be enriching its connotation, and rendering it all the more useful for later employment of it in other contexts. I should perhaps at this early stage answer a possible objection. It is that whereas most of the perfections listed by Barth have a fully biblical derivation, the notion of imagination does not appear in such a setting whatsoever in the Bible, whereas we saw at the opening of our discussion, it refers only to a human condition, and on the whole, an evil one at that. Two answers are possible to that criticism. First, theology has not in its history been tied to biblical terms in its exposition of its subject-matter, as is seen in its use of such terms as 'incarnation', 'history', 'Trinity' and *hypostasis*; secondly, Barth himself in this volume (II/1), as often elsewhere in his writings, uses the Latin concept of *aseitas*, of which the English word 'freedom' is taken to be one translation; and I would wonder whether even the word 'freedom' is attributed to God in the Bible in quite the sense in which Barth uses it in speaking of it in his review of the divine perfections. I should like now to draw attention to those parts in Barth's exposition where, to my mind, the notion of imagination could be introduced, or where alternatively it is implied.

First, Barth's whole emphasis upon 'the being of God as the one who loves', together with the central place which he gives to love in the account of the perfections, makes the relating of the imagination to love a natural starting-place. We follow Barth's repeated warning about the other attributes, namely, that we allow their application to God to be their logically primary meaning, otherwise we shall be engaged only in a general discussion of the connotation of terms. We shall endeavour at every point to ensure that the divine application will inform our concept of imagination, and not allow the reverse to happen. In this same way Barth had said (*id.*, p.276) we do not begin with a definition of the word 'love', but with the resolve to let the act of God visible in his revelation speak for itself, and allow the word to take on a meaning which is fulfilled in a way which breaks up and reforms its usual meaning. I would ask no more for 'imagination'.

At p.278 (*op. cit.*), when he develops his account of God's loving, Barth says that it is concerned with seeking and creating fellowship without any reference to an existing attitude or worthiness on the part of the person concerned, the point made so forcefully by Anders Nygren in *Agape and Eros*. I would detect in God's loving that feature observed in Christ's attitude to Zacchaeus, in whom there was nothing that might especially have endeared him to Christ. The love of God, as was the case in that love of Christ for Zacchaeus, shows as one of its special ways of working an imaginative penetration into the plight and condition of the loved one, worthless as he is. To describe this imaginative penetration by the loving God into the world of the sinner, Barth employs a range of images to fill out the concept: God's love is a light shining out into the darkness; it throws a bridge over the crevasse; it senses alienation and hostility, but it seeks to create fellowship with the abandoned one.

Secondly, at p.313 (*op. cit.*), Barth develops the theme farther, giving still fuller content to what may fairly be understood as the concept of imagination in its relation to the freedom of God. God's freedom, we are told, may express itself in his being present with that which is not God, in communicating himself in a way which entirely surpasses all that can be effected by way of reciprocal presence, communion and fellowship between other beings. No created being can be immanent in, or inwardly present to, other beings; but God in his absoluteness can both transcend all that is not himself, and also if he wills, be immanent within it. Barth is making the point that the immanence which God effects does not endanger his transcendence, or lead to any kind of pantheism or panentheism. Once again I would like to interpret this projection by God of himself into the state and condition of the other who is the sinner as an imaginative activity, based upon his deep understanding of, and sympathy for, the other. In a sense, it is the imaginative projection by God of himself into the other which is the earlier and necessary condition of the immanence. Such an interpretation I see as the implementation of George MacDonald's proposal that we seek ways of presenting imagination as one of God's attributes. Let me say, however, that in order to make out his case for the freedom of God, Barth has somewhat overdrawn the difference between God and ourselves. He says explicitly that it is of the essence of created beings that they should have frontiers which they defend jealously against the invasion of other personalities; nor can created beings be inwardly present to one another.

Created beings have to be true to themselves, and in so doing they can not be true to another. They affirm themselves only by doing so against others. Now I can see that Barth in these sentences may have unredeemed people in mind, though even among them there are in history and in contemporary happenings, far too many accounts of self-sacrifice to make such a view altogether credible. Certainly, as applied to redeemed men and women they must surely be false. The command, 'Love one another as I have loved you', must allow for — and indeed requires — a measure of imaginative penetration by human beings beyond the frontiers which they often set up to keep out not only God but one another. In fact, Barth would not have reduced the force of what he was saying about God's freedom and his ability to transcend other beings and yet to indwell them, by allowing something of that same imaginative penetration to be possible for redeemed persons. But there is another particularly valuable point in Barth's account of God's transcendence of others and of his immanence within them: it is that in this process of indwelling through imaginative penetration, God does not cease to be himself, nor does he turn into the other. The same would hold true of the imaginative penetration which we are claiming for Christian believers who seek imaginatively to indwell the misery and sufferings of their fellows. In so doing, they do not lose the identity of their own personality; they remain themselves, albeit greatly enriched through their experiences. This subject will arise later when in the context of the place of imagination in morality and ethics (*vide infra*, pp.000-000) we come to examine how the concept of identification works.

Thirdly, many of these same ideas recur when Barth passes on to write about the mercy of God (*op. cit.*, pp.369ff.). The love of God bearing necessarily the character of grace comprehends also the character of mercy. The free inclination of God to his creatures presupposes that the creatures are in distress, and that it is God's intention to espouse their cause and to help them in their distress. 'The mercy of God lies in his readiness to share in sympathy the distress of another', a sympathy which springs from his inmost being. God participates in the distress through sympathy, and is thereby present in its midst, and taking the initiative to remove it. At p.370

(*ibid.*) he criticises Schleiermacher for trying to eject the idea of mercy from the language of dogmatics and to transfer it to homiletics and poetry: 'the source of the feeling of sheer dependence has no heart. But the personal God has a heart. He can feel, and be affected'. My conviction is that when we have reached the point of using such language about God — and Barth does so with all reverence — we have achieved the possibility, or even, have accepted the obligation, to include in our description of God in terms of his attributes, that of imagination, which carries all these notions of penetration into the distress of another, sympathising with that distress, and providing the motivation for the relief of that distress. There is a sense in which it is imagination which initiates the several other activities and sustains them, through to the point where the relief of the distress is complete.

Fourthly, we obtain further amplification of the way in which the concept of imagination may be applied to the understanding of how God's love acts in the freedom of his being, when we come to Barth's treatment of the perfection called 'patience'. This perfection he regards as an enrichment, clarification and intensification of the idea of mercy, insofar as God takes up the cause of his creatures, 'realities distinct from himself', affirms that reality and does not suspend it. He takes the plight of the creatures into his heart, but without in any way depriving them of their independence. In fact, in this process, God's patience renews and transforms the nature of the creatures. As the passage on the patience of God progresses, Barth brings out clearly the ground on which God's patience finally rests — the One who stood in the place of the many, and who has secured for them time and space to repent. For our present purposes of the connection with imagination, we shall take up only the first point, namely, that God's mercy affirms the independence of the others, even in the moment of taking their plight to his own heart in patience. There is sensitivity of the highest order, which even in offering renewal of being, with all the rejection of the old being that that involves, can yet succeed in endowing the persons so renewed, with fresh existence as objects of God's patient mercy, living under God's righteousness, and enjoying to the full the outworking of the encounter with God. This expression involves no violation of the integrity of other persons, but only the possibility of their renewal and transformation at the hands of a God who suffers long and is kind.

Fifthly, if I were, therefore, to gather together the elements in the account of imagination as an attribute of God, as we might form it with the assistance of Barth's examination of the perfections of God, I would propose the following: (*a*) The imagination of God is a direct derivative of the love of God, which therefore acts as a control upon all that we say concerning imagination. Equally, it gives imagination access to the whole range of the actions of the God who is loving. (*b*) Imagination is the medium of God's loving penetration into the world of the sinner. It is the form which God's awareness takes of the condition of the other who stands over against him in rebellion and hatred. (*c*) Consequently, it is the ground of God's immanence in the plight of the sinner, immanence which nevertheless does not entail any loss on God's part of his transcendence of all created reality. (*d*) From God's heart imaginatively stirred to take the initiative towards making good the state of the distressed comes the effective action whereby that end is achieved. (*e*) But this process sensitively and imaginatively affirms and does not deny the independence of sinners, but patiently waits for their return to the Lord. The One who has stood in the sinners' place has given them time and space in which to come to penitence, the One who has come out of the Father's heart of love. Before leaving Barth's treatment of the perfections of God, which we have seen to be most helpful in enabling us to find a place for imagination in the description of God's attributes, I should like to enter a query concerning his handling of the subject. Clearly he has dealt with most of the attributes encountered in any traditional discussion of them, with the possible exception of omniscience, (which in any case might be subsumed under 'wisdom' in Barth's list), and he has done so in fresh and novel style. What I did find odd was the curious position in which he places love and freedom, which he does not classify as perfections or attributes, but which he frequently employs as adjectives when speaking of God, so that effectively they operate as attributes most of the time. He places them in such close relation to God as to be of the essence of his being as act.

He says again that they supply us with the two directions in which we may explore the exposition of the attributes, so that the lists which he gives amount to the working out of the meaning of God's love and freedom. But even such explanatory exegesis is not altogether satisfactory, in that when he formulates his theses on the perfections, he says not that 'the love of God consists in his being gracious, merciful and patient', but that 'the divinity of the love of God consists and confirms itself in the fact that in himself and in all his works God is gracious, merciful and patient'. The impression left by this rather wordy sentence is that it is God's being gracious, merciful and patient, which constitutes the evidence either for our saying that he is loving; or, more obscurely, for the divinity of God's love. The latter alternative I find singularly unrevealing; but I hasten to add that Barth's actual treatment of the attributes is considerably more revealing than the summary text of his titular propositions would lead us to expect.

Imagination and Creation

A traditional response to the request for a brief account of the doctrine of creation would probably yield the three-fold formula: creation is *creatio ex nihilo*; it is *creatio per Verbum* and *creatio continua*. By the first of these assertions we would be distinguishing the creative acts of God from all those activities in which we 'create' things, such as works of art, from previously existent materials. It also distinguishes Christian belief from those philosophies which regard matter as being co-eternal with God, and therefore as constituting a limitation upon him and his freedom. In the formula, the term *nihil* does not refer to a something from which God created the universe, as if by giving it a name we endowed it with some measure of reality. The formula intends that there was nothing from which God might have created the universe; he did not create it out of anything. The second phrase, *creatio per Verbum*, picks up almost literally, the language of the Prologue to St John's Gospel (St Jo 1.3b), 'he it was (the *Logos* or *Verbum*) by whom all things were made, and without him was not anything made that was made'. Originating in the Prologue, this formula has found its way into subsequent western and eastern theology. It is thought also to reflect what the early passages in Genesis say about the manner of the creative act. God effected the creation of the world and all that is in it, by simply speaking the words, 'Let there be ...', so that the creation was by word, *per Verbum*, and not by act. *Creatio continua* is the rejection of the position which was later to be known as Deism, the view that when he first created the world, God left it to its own devices, and withdrew into the solitude of his own being. On the contrary, says this third element in the doctrine, God constantly continues his work of creation.

This account of the doctrine of creation is one of impeccable doctrinal propriety, heavily loaded as it is with the correct Latinisms. Yet if we sense a certain vacuity in its presentation, we may have a certain sympathy with Barth who commented (*Church Dogmatics*, II/1, p.76, ET), 'within the sphere of ideas possible to us, *creatio ex nihilo* can appear only as an absurdity ... We have no analogy on the basis of which the nature and being of God as Creator can be accessible to us'. Clearly there must be something wrong with these doctrinally impeccable formulae if they are in fact another way of saying nothing at all. Without claiming that false accessibility to God which Barth abjures, we could perhaps attempt another way of approaching the matter. My question would be: could we not have employed language and references which did not fall with quite such a leaden thud as the imponderable Latin tones? When we speak of creativity in the human situation, the first notion that comes to mind in close relation to it is that of imagination. In fact, they are almost synonymous. Why then should we hesitate to extend the use of the term from human to divine creativity, and say that God's creative action is consummately imaginative? The suggestion is not so much to supplant the others as to supplement them and link them with that activity of mankind which bears them closest relation and resemblance.

Indeed there is no cause for us to be apologetic about the proposal. Nature red of tooth and claw must not obscure for us the beauty that there is for us in this world of God's creating. There is no reason why we should not, and every good reason why we should, see the brilliance of an Australian sunset in a sky of translucent opal, the unfolding glory of a Superstar rose, the thousand glows of yellow, golden, auburn and brown in a New England forest in the Fall, the awe-inspiring majesty of Mount Everest in the few moments before the last light fades from the sky, these and a thousand other glories that the poets have sung, as the expression of the imagination of God in his own creation. This is the kind of God with whom we have to do, who spreads forth beauty in such lavish profusion. Our long-sustained strictures on natural theology, our almost pathological unwillingness to allow any way from the world about us to the God who created it, have together almost succeeded in killing within us any perception of the character of God as it is revealed in the beauty about us. There surely is an inescapable truth in St Thomas's words that God is known *per ea quae facta sunt*, through the things which he has made, even though we do not follow him, as he intended us to do, down one or other of his Five Ways of proving that God exists. My counsel therefore is that we do not allow our understanding of the doctrine of God the Creator, who creates so imaginatively, ever to be entangled with our suspicions about natural theology: that is another theological subject, and it must be dealt with at another time. Nor am I arguing that we can ever fully know the doctrine of creation without an understanding of the doctrine of reconciliation. That may be so, but we have to find a place for the true appreciation of nature even after it is renewed by grace.

Protestant theology has been aware for some time of the inadequacy of its theology of nature. Two sources of this inadequacy are normally mentioned. One we have just noted, namely, its inborn suspicion of natural theology, with its suggestion not only of arguing from nature to God, but also of finally making God continuous with nature, as one cause, even though the first, among other co-ordinate causes. The second has been the much condemned doctrine of the *dominium* of man over nature, man's so-thought God-given right to use the resources of nature to serve his own ends. The result has been the secularisation of nature, with the way being left open for the spoliation and destruction which our generation is doing so much, though often ineffectually, to check. Two conditions will have to be met if the aims of such a group as 'the Friends of the Earth' are to be achieved. First, there has to be a quite massive motivation operating at all levels — commercial, industrial, social, economic, political, public and private, moral and religious — to contain the pollution that is still perpetrated in our midst and around us, in earth and sea and sky, and, despite all our determined efforts, perpetrated by ourselves. We are still using one form of the greed motive to limit another form of it. The appeal on behalf of generations yet unborn carries very little weight with a generation which believes that it must despoil in order to maintain its standard of living; while, in any case, altruism has never proved to be a very potent bargaining counter in disagreements between nations, notably for example, in relation to fishing limits. The second requirement for a successful attack on the ecological issues of our day is a convincing philosophy you would call it, if you are not religiously inclined — some picture of the world as you would like to see it be, a picture that would begin to move and to attract those who normally see no farther than the half-yearly company report. But if you come to those matters with a Christian interest and obligation, then what, I have to say, the situation requires is a theology of nature which is sufficiently earthed (no pun intended) to be relevant to what is one of the most pressing issues of our day. Our first step is to counter the secularisation endemic to our tradition, re-discovering God's own involvement in nature. As said before, I have no intention of identifying God with nature: the doctrine of God the Creator should by now be preventing us from committing such an error. But we have to begin again to see nature as the Psalmist saw it, the handiwork of God which our ineptitude and greed are destroying; or as Christ saw it, with God clothing the lilies of the field in a glory which outshone Solomon's finery, and having such a care for the minutiae of nature that he was aware of a sparrow falling to the ground, and, by implication, of ten thousand gannets destroyed by one carelessly released oil-slick. Somewhere in the midst of that theology I want to find a place for the imagination of the Creator-God, so that our theologico-ecology does not confine itself to the utilitarianism of resource-conservation, laudable as that is, but is comprehensive enough to realise that we have to preserve the beauty of the world so imaginatively and profusely loaded with such splendour by God. But I may be in error in separating the two, for all too often the resource-destruction is a beauty-destruction. Conserve the one and you conserve the other. My conviction is that we shall require to mobilise the entire forces of our theology if we are to make any inroads into the ecological tragedies of our time. So far we have dispatched one single doctrine to hold the line, and understandably the Christians at the front facing the despoilers and the resource-destroyers complain that we are not supporting them. But the first step in mobilising the whole of theology is to get a right beginning in our doctrine of nature and of the God who so imaginatively created it.

Imagination in the Incarnation

I should like to introduce this part of our study by looking at the parable of the dishonest tenants (Mt 21.33-45; Mk 12. 1-12; Lk 20.9-19). A householder let out his vineyard to tenants, and went abroad. When the time came to harvest the fruit, he sent his servants to collect his dues. But the tenants killed them. He sent more servants, but they met with the same fate. Finally, he sent his own son, believing that they would respect him. However, the dishonest tenants now saw an opportunity to seize the inheritance for themselves, so they killed the son. Those who heard the parable believed that the owner of the vineyard should now punish the wicked tenants with death, and give it to others to rent. The end of the parable is eschatologically oriented, for in the explanatory words which follow, Jesus warns that the Kingdom will be taken from those who refused to receive the Son when he came, sent of the Father, and be given to others. It is the earlier part of the parable which is my concern, since it may be taken as a parabolic account of how the Incarnation came to be. God had sent his prophets, declaring in generation after generation the will of God for his people, making plain his goodwill towards them and his high purposes for them, promising that should they repent of their sins and return to him, he will still forgive them. Yet in generation after generation they stoned and killed the prophets. Even then God did not abandon them. In a single act of what can with reverence be called daring imaginativeness, God resolved to send his Son in the form of a servant, the Word made flesh, yet a man among his fellows. The message, the Word, which somehow they had been unable to grasp when spoken to them, was bodied forth, in flesh and blood, to be seen and heard, touched and handled, in a medium and in terms unmistakeable; now, they would be without excuse. When the phrase 'daring imaginativeness' is applied to the Incarnation, it is no exaggeration to say that it is barely acceptable to most of us. It comes as a kind of negative cultural shock, or should we say, as a cultural non-shock. It is so for two reasons. First, we have heard the Gospel story so often that familiarity with it has removed most of its novelty for us. Such familiarity makes it virtually impossible for us to think ourselves back into the original situation of Christ coming among men and women, and virtually impossible for us to grasp what Kierkegaard meant when he spoke of our being contemporary disciples of Christ. If we ever do make the effort, we rather anachronistically carry our familiarity with us, and imagine that we would perceptively have beheld in Jesus Christ the Son of the Father. We would not have been numbered with those scribes and pharisees, whose religion blinded them to the truth which was open for all to see. In fact, there was a great deal of credibility about the reaction of the scribes and the pharisees; at least, they did appreciate the sheer novelty of what was claimed to be happening, even though they denounced it as blasphemy. As contemporary disciples, we would probably have been no more aware of the true character of Jesus than the disciples originally were! As each Christmas returns, the greatest single responsibility which Christians bear is to try to recapture something of the sheer unexpectedness, the stunning imaginativeness of that more excellent way which God finally chose, to make his love and forgiveness unmistakeably real to his people.

The second circumstance which makes it so difficult for us to discern the daring and totally unexpected imaginativeness of the Incarnation is the central place which the argument from prophecy held in the *kerygma* of the young Church. The intention of that argument was twofold: first, to demonstrate that Jesus was the deliverer that Israel had been promised by God from of old; and secondly, to derive from the prophecies the terms in which to interpret the person and mission of Jesus, very much as according to the Synoptists Jesus himself used Isaiah 53 to interpret his death. But an unintended consequence of the wide use of the argument was that it introduced an almost logical necessity operating between the prophecy and its fulfilment, and thereby eliminating the element of novelty and unexpectedness from the event of the Incarnation itself. Once again hindsight, with the fulfilment of the prophecy before it, selected from the prophecies those features of them which were in the event fulfilled, and tended to neglect those which were not. Prior to the event, it would be extremely difficult to predict which elements in the prophecies were the most likely to be fulfilled, as the wide range of messianic movements found in Israel around the end of the pre-Christian period and into it demonstrates. Further, that distinctive feature of the fulfilment of prophecy in Christ, what we call the Incarnation, the presence of God in person in the midst of mankind and living a human life, is totally absent from the prophecies of the Old Testament. There has to be a gap of discontinuity between prophecy and fulfilment; there is no point-to-point correlation between the two, so that if you have the one you can predict the other. The unexpected, the unpredictable, stands in that gap; and it is there that the imagination of God does that which far exceeds our aspirations or our deserts. That imaginative creativity which God showed in forming the beauty of the world about us, he demonstrated once again in the novel style of his intervention in human history in the form of the Word made flesh, an event so unexpected, so unpredictable, that those who might have done so because they were of his chosen company, failed to grasp the full wonder of what was happening before their eyes; while those who had read the signs and oracles, who knew their Scriptures and were daily expecting a Messiah, brought him out to Golgotha to ensure that their rejection of him was total. I know of no better argument for placing imagination at the heart of God's dealings with us than the single, unique, unpredicted and unpredictable event of the Incarnation.

56   FAITH THEOLOGY AND IMAGINATION

Imagination in the Atonement

In our general strictures on the aniconastic strain which has characterised most of the history of theology, we should have given notice that there is one area of doctrine and theology where this criticism was not true, namely, the doctrine of the atonement. Why this fact should be so is a matter to which we shall return. Certainly the use of images, and by implication, of imagination, is almost universal in accounts of the atonement, as may be evidenced by a review of images employed in descriptions of the death of Christ. Sometimes the term 'model' is used in this connection, and I may follow the practice, though I would be using the terms 'model' and 'image' interchangeably, and recognising that there is a very interesting discussion as to how the two might be related. Histories of soteriology (e.g., T. H. Hughes, *The Atonement*) leave you with the distinct impression that these different images or models have grown up over the centuries, and have come to be identified with the names of specific theologians in, say, the twelfth or the nineteenth centuries; whereas, in fact, they are most of them original to the Primitive Church. On the other hand, some writers, such as Emil Brunner (*The Mediator*), prefer to concentrate on one or two of the images to the exclusion of the others. Here, then, is the list: (*a*) The image of *ransom* seems to be among the most original employed to describe the death of Christ, appearing at Mk 10.45 in the words of Jesus, 'For the Son of Man also came not to be served but to serve and to give his life a ransom for many'. The image carries with it the notion of release from prison or bondage on the payment of a sum; what the image does not indicate is to whom the sum is to be paid. Early soteriology tried to complete the image by saying that the payment was made to the devil; but such a suggestion is intolerable, though it was a thousand years before it was exposed as gross theological error by St Anselm, and even then did not disappear from popular thought for centuries after that. The image, in other words, has to be considered as an 'incomplete symbol', a point that appears in other cases. (*b*) From II Cor 5.19, 'God was in Christ reconciling the world to himself, not counting their trespasses against them' comes the model of *reconciliation*. Once again the model is not operating totally in accord with its own structure, or implying what we would justifiably expect it to. It was mankind who had offended God by wrong-doing, and not *vice versa*, and therefore we would expect that God, being the person offended and alienated, should be the one to be reconciled. Yet the model operates the reverse way, and God is said to be reconciling *us* to himself. (*c*) There follow two images which are close to one another, and both have strong Old Testament connections. The first is *sacrifice* (Heb 10.10, 12: 'the body of Christ, a single sacrifice offered for all time for sins') which links with the Old Testament sacrificial system, but is presented as bringing it to an end. Again we have an incomplete symbol, for it is not clear to whom the sacrifice is offered, certainly not to God if it is suggested that he has to be appeased. (*d*) The closely connected image is that of *atonement*, which carries similar connotations of a subject requiring to be appeased, though that idea is clearly not part of the image-pattern which this soteriological statement accepts. The link is with the Old Testament Day of Atonement, and the image indicates that Christ is the one who has carried away our sins, and made peace on our behalf with God. (*e*) *Propitiation* is not far in meaning from the two previous images. St John speaks of Christ being the propitiation for the sins of the whole world (1 Jo 2.2). The completion of the symbol is to be avoided, if by completing it we are drawn to say that God (conceived of as an angry eastern potentate) has to be propitiated because of the defiance of him in sin. (*f*) We come now to *salvation*, the image which answers the question, 'from what?', with the words, 'from sin, from the devil, and from the powers of darkness'. It is supported by the description of Christ as 'Saviour'. (*g*) *Redemption*, another very widely used image, contains something of the original ransom notion, but it appears sufficiently frequently in the Bible and in soteriology on its own, to be classified separately. It answers the question, 'from what?' very much in the same manner as did the image of salvation. (*h*) The image which Gustav Aulen regarded as absolutely primal in Christian soteriology was enshrined in what he affirmed to be the original Christian confession, *Christus Victor*. In his death Christ vanquished all the principalities and powers, the rulers of the darkness of this world, who had long held mankind in their thrall. There was no question of offering them ransom; they were totally destroyed. (*i*) The image of *satisfaction*, which carries the connotation of meeting or satisfying the demands of a righteous and just God, had several forms of presentation: in the *Cur Deus-homo* of St Anselm, satisfaction has some feudal overtones, and the satisfaction which has to be made has to relate not so much to the deed which failed to be done because of the disobedience of the sinner, but rather to the amends to be made for the affront to God's glory implicit in the disobedience. That satisfaction, which man ought to offer but can not, and which God ought not to offer but can, is in his death achieved by Jesus Christ who is the *Deus-homo*, the God-man. Another form of the satisfaction image employs what is effectively a second, qualifying image, namely, punishment, to give the composite image of 'penal satisfaction'. In this case the satisfaction is achieved by Christ, who in his body endures the penalty of suffering and death, which is God's judgment on sin and the sinner. A third form of the satisfaction model, designated 'moral satisfaction' (later to be identified as 'vicarious penitence'), describes the way in which Christ by his moral actions satisfies the demands of the wholly good God, in a way which is impossible for sinners, and offers these righteous actions (or in the other case, offers a totally adequate repentance for sin) on behalf of his sinful brethren who thereby become acceptable to God. (*j*) We must, however, mention separately *vicarious penitence*, because of the outstanding work of John McLeod Campbell (*The Nature of the Atonement*). God requires of sinners repentance in like measure to the enormity of their sins, which only Christ can achieve; and he does so by identifying himself with them in their sinful condition. (*k*) It would be wrong to omit the model of *moral example*, which has been so much promoted by those who either found the so-called objective theories too commercial, or too sub-personal, or felt that they operated with a pre-Christian view of God. They might take as their starting-point such words as St Paul's in Phil 2.5: 'Let this mind be in you which was also in Christ Jesus'. The self-sacrifice of Christ then becomes normative for his disciples. (*l*) A final model which has acquired immense popularity in modern times is that of *revelation*. The death of

Christ is seen to be the supreme act of self-revelation by God, in which he makes known to his people his purpose of salvation for them. This is what might be called a 'second-order' model, for it does not of itself indicate what exactly it is that God reveals, or how he reveals it, without employing some of the other images, his purpose then being defined as that of reconciling his people to himself, or redeeming them, and so on. So much then for the images which are basic to the different theories of the atonement which have appeared in the history of theology; and it is noted again that despite the degree of conceptualisation involved in such theories, the vast range of images which we have listed have refused stubbornly to go away. But the provision and entertainment of images is not the sole role which imagination plays in the doctrine of the atonement and we observe in passing how that one image has come to stand stead for all the others. Another special feature of this doctrine, as it has been variously stated in biblical and theological literature, is that it also provides us with what I call 'relaters', which are further images designed to relate the soteriological images, or more exactly, the situations which they describe and define, to the persons for whom the salvation, the redemption, and so on, were designed. Alternatively, these relaters relate the One to the Many, qualifying the model in favour of the Many, and employing what would appear to be another set of images to do so. These relaters are as follows: the first is the image of 'substitute' giving the adjective 'substitutionary', and it draws attention to the fact that what Christ does, he does as a substitute for ourselves, 'in our place', *anti hemon*. What Christ did he did in our place, dying the death which we by reason of our sin deserved to die. 'Vicarious' would be an alternative relater. Another relater which has been widely used in soteriology is 'representative', the idea this time being that what Christ did, he did as representing sinners. In receiving the punishment which sin deserved at the hands of the righteous God, He died as the representative of all those whom he called brothers and sisters. A further relater, which is still so even although it comes in the form of a phrase, is 'on our behalf'. It is slightly different from the first two, and indicates that what Christ did for us, is transferred to us by God, in particular, the cancellation of our sins. The final relater which I wish to mention, another description of how we are integrated with the work of Christ on the cross, is that of 'identification'. The *locus classicus* for this model is II Cor 5.21: 'Him who knew no sin, God made to be sin for us'. Christ was identified with us in our alienation from God without himself becoming 'sinner'. The consequence for us is that we are become his righteousness. The single, unique, once-for-all act of Christ universalised through the relater, his being identified with us. We shall conclude with a few comments, drawing out some further features of the role of images and imagination in the construction of theories of the atonement. First, it has on occasion been said that the theories of the atonement if worked to their logical conclusions would make it appear that forgiveness on God's part was no longer either necessary or possible. All had been done to meet the requirements of his righteousness, according to the theories, without his having to offer to forgive. However, as we have seen, the images and models employed in atonement theories are 'incomplete symbols'; they do not follow their own structures to their rigidly logical conclusions. Rather ought we to see the theories with their various images, not so much arguing a certain logic, as portraying to us, through these images, the way in which forgiveness works. They represent the processes initiated by God to give effect to forgiveness. Secondly, the images and models of the theories of atonement indicate also how forgiveness is appropriated. We insert ourselves into the structure of one or other of these images or models, and in so doing avail ourselves of the ransom which Christ has paid for our sin, or claiming a share of the victory which he has achieved over the forces that debilitate us, or being reconciled to God through Jesus Christ, and so on. But the uniformly effective point is that we come to forgiveness by means of one or other of these models or images, which therefore are severally the paths to salvation. Thirdly, they represent the terms in which we proclaim the Gospel of salvation by Jesus Christ, the media in and through which we present to others the story which will bring them to Christ. They have, therefore, a kerygmatic as well as an expository function to fulfil. It is this purpose which has in the end saved the doctrine of the atonement from the invasion of aniconastic thought. It lay too close to the heart of Christianity to be conceptualised into ineffectuality.

Imagination and the Holy Spirit

Thirty years ago there were wide complaints, some of them echoes of what had been being said for the previous fifty years, that neither had we an adequate doctrine of the Holy Spirit, nor did we have any clear understanding of how the Holy Spirit works. Neither of these complaints would be countenanced today. We are able to give competent answers in both divisions, but now we have another problem, namely, is there any single uniting principle which will hold these seemingly disparate and heterogeneous phenomena of our time — a reformulated doctrine of the Holy Spirit, and an almost totally unprecedented outburst of Holy Spirit expression and activity? Tentatively, because the very idea will be most unfamiliar to those who are acquainted with orthodox formulations, I would like to examine the possibility that the notion of imagination, functioning through images and models, might offer grounds for such integration. Before exploring that answer, let us see how the two sets of phenomena have defined themselves.

On the one hand, the doctrine of the Holy Spirit has in conservative theological circles been emerging along certain clear lines, anticipated some fifty years ago by Karl Barth (*The Doctrine of the Word of God*, ET, pp.513ff.) thus. First, adopting the popular modern emphasis upon revelation (and very few contemporary theologians of whatever persuasion fail to do so), the doctrine as now defined says that the Holy Spirit is God's guarantee that revelation will be revelation. Given the radical sin in humanity, then even with God revealing himself, he had no assurance that what he had made known of himself for human salvation would ever get through to its destination in human hearts. The whole process might be stultified at the point where it was due to do most good. So rather than take the risk, God sent his Holy Spirit into the hearts of men and women, in order to guarantee that what has been revealed will be fully understood, that is, that revelation will be, as we said, truly revelation. Secondly, in much the same way, it is held that the Holy Spirit is God's making himself sure of us. Left to our own devices we could never be relied upon to receive salvation when it was offered to us by a gracious God. So God has to take steps to ensure that the salvation effected in Jesus Christ reaches the goal God had when before the foundation of the world he planned that salvation, namely, our here-and-now redemption. To make sure of that, and to make sure of us, God gives us the gift of the Holy Spirit. Another way of putting these first two points is to say that the Holy Spirit is God from above meeting God from above, the God from within the sinful heart meeting the transcendant God coming down from heaven to be enfleshed, to suffer, to die and to be raised again, for the saving of men and women. When these two meet, salvation is effected. In each of these ways, then, it is to be understood that the Holy Spirit is truly God, the third person of the Holy Trinity. His action in each of the three ways above-mentioned is what the old theologians used to call an 'appropriation', an assigning to one of the three Persons of the Trinity some particular work in relation to the created order (an *opus ad extra*), in which nevertheless all three Persons must also be thought to be participating. In this way, these varying accounts of the Holy Spirit are not to be thought to support a tritheistic interpretation of the Trinity.

The connection which I find between this modern account of the activity of the Holy Spirit within the economy of salvation and the notion of imagination lies in three directions: first, of the projection of God into the situation of sinners, removing the blindness which prevents them from perceiving the revelation, and the sinfulness which resists God's grace; secondly, God's involvement with all the minutiae of the condition of sinners, in a manner of high sensitivity and perceptiveness; and, thirdly, God does so, the transcendent becomes immanent, without thereby losing his transcendence or being transformed into the other. These three forms of the Spirit's activity are especially characteristic of the imagination, as we saw when discussing the attributes of God (*vide supra*, p.48).

On the other hand, the outstanding feature of the contemporary interest in the Holy Spirit is the widespread

A THEOLOGICAL CATEGORY 63

recognition throughout the Church and beyond it, of the manifest activity of the Holy Spirit, not only in the Pentecostalist and Neo-Pentecostalist Churches, but in the charismatic movement in the mainline churches. The evidence is common knowledge. Speaking with tongues has become a phenomenon more common perhaps than at any time since the Primitive Church, and now more sought after by some earnest believers than ever before as a mark of true faith. Divine healing now generally accepted by the established churches world-wide, has brought a new appreciation of the continuing healing ministry of Our Lord through his Spirit. It is almost as if, whereas an earlier theology linked the Holy Spirit almost entirely to the task of sealing the benefits of the saving work of Christ upon believers, now, in addition to that salvific work, the Holy Spirit is also linked to making effectual that healing ministry which was so conspicuous a part of his earthly ministry. There is, too, that ever growing phenomenon of baptism by the Holy Spirit, held by those who experience it to be the only final confirmation of discipleship. There is, I find, a good deal of pressure from very tidy-minded theologians to reduce these many different happenings to some semblance of theological law and order; and their anxiety is understandable, that we should not in our day of ecumenical drawing-together be giving encouragement to the multiplication of sects. But perhaps three responses might be made to this overcautionary word. First, the charismatic movement, or the Spirit movement as we may prefer to name it, appears in most of its forms in all of the major denominations, and therefore, so far from being a force for fragmentation, it seems to have the power to transcend most denominational barriers. Indeed, there are some who will go as far as to say that it is in the new unity in the Spirit that the ecumenical movement will gather fresh impetus as this century draws to its close. On the human side, the movement meets an uneasiness about a Church which is sometimes in danger of identifying good order and structural unity with the true life and being of the people of God. Secondly, we have, I feel, to follow St Paul in our approach to, and assessment of, the movement when he said that he had one test, and one test only, to put to the utterances of the Spirit, namely whether they declared that Jesus was

Lord; and if they did not they were to be condemned. Such a test gives a wide range of tolerance to charismatic activity, and should reduce some of the lack of appreciation which has at times marred organised Christian attitudes to the movement. In our previous section, we were outlining some characteristic forms of imaginative activity in what the Holy Spirit is discerned as doing, in the view of contemporary theology. I propose that the charismatic movement is yet another demonstration of the Spirit's acting imaginatively. Here we have a facet of imaginative activity which is not controlled and conditioned as some of the others were. In fact, we have something not dissimilar to the riot of extravagance which we witnessed in God's imaginative creativity in the sunset, or the magnificence of the rose, or the majestic fury of a storm at sea. There is a lavish richness about it all which can not be contained within the tidy categories of a suburban aesthetic. So the imaginative creativity of the Spirit in the extravagance of the charismatic movement defies categorisation by any over-neat theological pigeon-holer. We may do well not to fight against the movement lest we be found to fight against God himself.

A similar construction might well be put upon the events of the original Pentecost. There are many points about the record of that occasion which baffle the critics, and which biblical theology has not been moved urgently to adopt, and has in fact passed over in silence. But we could conceive of that first Pentecost as the extravagant expression of God's imaginative creative activity in the spiritual sphere, diverse, uncoordinated, and confusing to the tidy mind.

To gather together finally these two very different accounts of how the Holy Spirit is described as being active in the Church and in God's people today, I would venture to say that the Holy Spirit is God's imagination let loose and working with all the freedom of God in the world, and in the lives, the words and actions, of the men and women of our time. What look like two disparate series of phenomena are in reality the different expressions of the one single Person of the Godhead, the Holy Spirit, whose characteristic activity, the *opus ad extra*, to which he is appropriated is imaginative creation in the spirits of believers, and of unbelievers in whom he works according to God's uncovenanted mercies.

CHAPTER 4

IMAGINATION
THE ETHICAL DIMENSION

Devotional Principle

By way of introduction to our next theme, I propose to analyse what appear to me to be the major tensions which exist in the field of ethical theory. Such analysis, it is hoped, will enable us the better to assess the contribution which the imaginative principle may make to this very vexed and difficult part of Christian theology. Because of the immediate practical implications of such a contribution, we shall see also the role which imagination has to play in the understanding of how the Christian life is to be lived.

The different tensions to which I have just referred are as follows. First, *universal law and situational idiosyncrasy*: on the one hand we appear to receive from the Bible at many points, the clear indication that there are commandments and statutes delivered to God's people, which are universally valid and imperatively categorical. In a sense, the story of the people of Israel can be traced by reference to the co-ordinates of obedience to, and violation of, the commandments of God. But such awareness of the universal applicability of these principles and commandments is not in any sense archaic: those of us who have had any form of Christian upbringing will testify to a deeply ingrained sensitivity to this same assumption of the binding character of the commandments of the Scripture. Nor is this quality reduced, however often the commandments are broken or their obligatoriness denied. Nor are they respecters of persons, for David as king was as subject to their judgment and their applicability as when he was a young shepherd boy. On the other hand, we have come to realise acutely in recent years that no two ethical situations are ever alike. They will differ in times, in components, in social pressures, in economic conditions, as in the persons involved in them, or the persons involved in them will have changed with the passage of time. Because of such idiosyncrasy, it is argued, it is rather rough justice to apply a single law or commandment and thus over-ride insensitively the features which make each situation peculiar, even unique. 'Circumstances alter cases' so that a judgment appropriate to one is said to be inappropriate to another. Also, without diminishing the universality of the law, we may draw attention to the fact that it is very difficult to decide to which general class of action a given proposal might belong. Such classification may depend on a large number of non-ethical though ethically relevant facts which would suggest inclusion in one class rather than another, according to the degree of accuracy and comprehensiveness of our knowledge of them. Further, when we are given groups of commandments to be the basis of our moral life, no indication is ever given of how we are to resolve cases of conflict between such commandments. 'Conflict of duties' is a well-known ethical embarrassment. Finally, it can not escape our notice that Jesus himself appears to question the omnivalidity of at least some of the established statutes of Israel, when he says, 'It has been said unto you of old, but I say unto you ...' Now admittedly he did so in very controlled situations, and in relation to only a few old statutes, but the fact that he did so at all destroys their long-standing inviolability. St Paul follows a similar course when he indicates that certain of the old regulations relating to food as well, more importantly, as to circumcision, would have to be abandoned.

Out of this tension has developed the so-called 'situation ethics', whose dilemma was clearly stated by Bonhoeffer (*Ethics*, p.22), 'The will of God is not a system of rules which is established from the outset; it is something new and different in every different situation in life, and for this reason a man must ever anew examine what the will of God may be'. This ethical theory has come under severe criticism wherever it has been carelessly and inadequately stated. For example, it is thought to imply that if in unusual circumstances an act can be shown to be in some sense justifiable, then a normal moral condemnation of it as bad is inapplicable; also that there is no absolute good and evil; and that it fails to realise that even though the final choice for us in some morally complex situation should lie between two evils, nevertheless the sequel may show that the greater of the two evils was chosen though that was not clear at the time of choosing. Despite all this very valid criticism that may be made of the varieties of the situation ethic which we have witnessed of late, we must observe that the problem of the tension which we are at the moment examining is not a latter-day phenomenon. The components of the tension were not unnoticed by Jesus Christ, and they were endemic to the ethics of the nation of Israel even before that. Last century Kierkegaard gave it its classical name in the modern period when he called it 'the teleological suspension of the ethical' in *Fear and Trembling*.

Secondly, the tension between *persons and principles*: this tension has appeared in recent times in the context of the previous one, but it is sufficiently important to merit separate treatment. It derives from a recognition that the primarily important component in any ethical situation is not the principle, but the person, the agent who is acting in it. Sometimes, so it is argued, the person has appeared to be sacrificed, even victimised, for the sake of some principle that was thought to be inviolable; the boy Isaac in the biblical episode which Kierkegaard uses so movingly in *Fear and Trembling*, would be a good example. As we develop our ethic of persons, supported as it has been by a sophisticated metaphysic of personhood, as in Martin Buber and John MacMurray, then the more we become aware of the uniqueness of persons, and of *their* inviolability. We come to appreciate their idiosyncrasies, the pressures under which they often have to act, the past history which limits the possibilities of action open to them, as well as the range of attitudes available to them. In other words, just as in the past a principle-oriented ethic was prepared to sacrifice persons to principles, so now a person-oriented ethic will find itself sacrificing principles for the sake of persons. In a short time there evolved and was popularised the saying, almost a slogan, 'Persons before principles', the purpose of which was to enunciate that personal relationships based on love and formed by love, are superior to principles which may on occasion attempt to override such personal relationships for the sake of some impersonal principle. This tendency has been furthered by some pastoral and counselling techniques which emphasise the importance of affirming the person, and the danger of the judgmental attitude implicit in talk about principles.

It would not be out of place to mention how close this particular tension comes to one which played a central part in ethical discussions some fifty years ago, the tension between the right and the good. These discussions dealt with the question of which of these two was the ultimate ethical category, and which was subordinate to the other. Was the good ultimate, a certain conception of the good life and the good person, the right deed being that which was the expression of the good character in any given situation? Or was the right ultimate, the categorical imperative in whichever of its forms, dictating the action which the rational person must do in the given situation, the good person being the one who has acted rightly in life's situations?

Thirdly, the tension of *the ideal law and the practical compromises*: this tension was, of course, a dominant feature in the ethics of Reinhold Niebuhr; and no one has ever succeeded in exorcising it from the field of ethical study, or from the realities of ethical action. The two elements in the tension are as follows. On the one hand, it is admitted that the ideals of an ethic, and particularly of the Christian ethic of love, are absolute. They contain within themselves no qualifying conditions or escape clauses. They make no concessions to the imperfections of human nature or to the pressures of circumstances. 'Thou shalt love thine enemies'; 'Thou shalt forgive to seventy times seven, and beyond' — no quarter given to human frailty and instability. On the other hand, such is human nature, and such the moral state of human society, social classes, economic groups, and national communities, that there is no immediate way in which the absolute ideals can be implemented in any of those areas, without serious diminution of their absoluteness, and therefore stultification. Some have attempted to resolve the tension by declaring outright that the absolute ideals are irrelevant to the sordid conditions of human existence at any other than the strictly private and personal level, and even there dubiously because of human imperfection and sin. Others have attempted to write the limiting conditions imposed by inadequate human nature or sinful society into the ideal itself, so that while it is retained as a criterion of human and social behaviour, it is presented in a sub-absolute form capable of implementation. Others, notwithstanding, have sought to apply the absolute naively and simplistically to human affairs. It was said that the literal adoption of the absolute and perfect law of love by pacifists in World War II was a form of this use of the ideal; whereas the affirmation of the so-called four absolutes by the MRA would be another. Niebuhr's own treatment of the tension was not to resolve it or diminish it or abandon it or yet try to apply it *simpliciter*. Rather, while holding to the absoluteness of the ideal he claimed that it was relevant in a number of ways; for example, it would indicate the direction in which agents ought to move if they are setting out to improve a situation which is to be condemned as immoral or unjust; moreover, it would provide them with motivation towards such action; it would prevent them from being satisfied with the compromise for which they might be tempted to settle, given the intractability of circumstances and their own imperfections; it would remind them that they ought to press on to the next stage of amelioration; but also, and importantly, it would prevent them from sinking into despair over the inapplicability of absolute ideals. Without going into the relative merits of these different ways of dealing with this third tension, we can see that the variety of these ways testifies to the reality of this tension as a major ethical problem.

Fourthly, we have the tension of *the divine imperative and our human responsibility to discover the will of God*. A great deal has been made in recent Reformed theology of the concept of the divine imperative, a title employed by Emil Brunner for his major book on ethics. The concept was not so used by Brunner himself, but not a few who were of his persuasion held the idea that the paradigm Christian ethical case was one where God decreed and the Christian obeyed, the decree coming *senkrecht von oben*, a particular word for a particular person in a particular place. In this respect, and despite the many other notable differences, such an ethic has a great deal in common with 'situation ethics' with the emphasis on particularity. The recognition of the important role of the will of God and his imperative can never be very far from the heart of Christian ethics, and Christian moral obedience. By and large, we hope that we are following God's command to us. But having said so, we hasten to acknowledge our uneasiness. As sinners, we can never be absolutely sure that we have heard God aright when he issued that imperative to us. We have observed also that people have received what they claimed were divine imperatives, and have committed themselves to totally opposed lines of action; one of the groups must be in error. Moreover, when we look at the actions which are said to be decreed by God they do fall into the fairly regular patterns indicated by the commandments of Scriptures, and not surprisingly so. So the other part of the tension emerges, namely, that the role of the Christian practitioner is not to sit waiting for some word from above, but to search the Scriptures diligently, in order to discern what that word might be. It is extremely significant that Karl Barth, who speaks so often in the Divine Imperative strain, when he comes to discuss the very vexed problems of contemporary ethics, such as suicide, the termination of pregnancy, and euthanasia, resorts to the deliberative method of ethical reflection, admitting that it may end in error of final judgment, but also that there is no alternative method in the form of a short-cut. So much of our contemporary ethical thinking not only in relation to the topics noted in Barth, but also such matters as the mining of uranium, political affairs in Northern Ireland, Southern Africa, Afghanistan, Poland, Ethiopia and El Salvador, and so on, has to be done in that piecemeal, enquiring way, that seems almost to be normal. The end-product seems almost to be no more that a potentially highly fallacious judgment, far-removed from the veridically declared divine imperative. Yet there the tension lies, for even the final decision we hope somehow is not unrelated to that imperative. The tension is too strong however for us to make a simple identification, and there the tension remains.

Fifthly, 'the good I would I do not; the evil I would not, that I do' describes a tension as old as morality itself, testifying as it does both to the moral aspiration of mankind and to their failure to achieve the best they know. It is the tension between

THE ETHICAL DIMENSION

God's standing offer of succouring grace, and our pathetic inability to accept it as a free gift. In recognising the fact of this tension, we are affirming that the ethical problem is not one simply of knowledge, of discovering the right action to do when required; but also of how we can do that action once we are aware of it. We may say simply that all that is necessary is that we open our hearts to that grace; but there precisely lies the problem of: how? The possibility that will be taken up later is the way in which imagination may help in the presentation of the right action to the believer, in order to increase motivation.

Sixthly, the question is always raised in traditional ethics about the tension between *freedom and necessity*, with the question of responsibility waiting in the background as the ultimate moral test of both. Writers such as Monod on freedom and chance, and Skinner on psychological behaviourism, have revived controversies which were dormant in that form for decades. But the popularity of the social sciences has made us all aware of the immense forces of social conditioning, economic pressure and political ideology of whatever colour, to which we are all constantly exposed, and which taken together seriously limit the range of free options that are genuinely available to us. Stated in this form, the problem is not simply one of how we are able to act with any freedom, and therefore with any degree of responsibility, in our gravely constricted circumstances, but also how the grace of God may have access sufficiently open to reach us. A prison-house shuts out in addition to shutting in, and the exclusion and the inclusion are the two faces of the same problem. In describing the matter thus, however, we have raised a further, a theological, form of the problem of freedom, namely, the problem of the relation of divine grace, sometimes described as irresistible, to human freedom. How can the one exist without stultifying or even destroying the other? If the one is destroyed by the other, how can they continue to exist side-by-side, so that we say with St Paul, 'I, yet not I, but the grace of God in me'?

Seventhly, our final tension is that between *authority and freedom* in the transmission of value in education. There is probably no more serious problem for our generation, and certainly no greater one for the Church in our generation in its relationship to the young, than that of how it can transmit to its heirs the values, standards and religious insights, which it itself accepts. The problem is not simply one of communication, of how A explains to B the content and meaning of an experience with which B has not yet had first-hand contact, and for which he or she has no desire. There is the question also of how you engender the concept of commitment as being essential to the proper attitude to values and to the religious Reality, without resort to authoritarianism. For many this tension constitutes the crisis of education, and particularly of Christian education, it even being said that we are the first generation which has not cared enough about its values and religious concepts, to make sure that they are effectually transmitted to the next generation. Instead we have substituted a pluralism which is intended to offer the young the range of choices among which they may make up their minds or not as they choose, failing to see that no decision can possibly be made among these choices unless the process of, and the grounds for, valid decision-making have also been imparted to them. Pluralistic presentation by itself could well be the first stage in a sequence which leads through subjectivism, indifferentism, agnosticism, atheism and finally to nihilism. There may be no logical entailment linking the several stages of this sequence, but there certainly are emotional and volitional relationships obtaining among them.

I do not propose now to deal with each of these tensions or problems in series, showing how imagination is relevant to all of them, and may help in our understanding of them. Rather shall we begin by indicating how imagination works as a principle in a number of major ethical concepts, most of which are involved in these tensions, later returning to pick up any loose ends which have dropped out in the analysis.

Let me begin with an examination of the part imagination plays in the central principle or concept of Christian ethics, namely, love. A number of years ago, I attempted to provide what could be called a structure for this principle in response to the recognition that it had become debased currency in much of the artificiality of television, films and some modern literature. My proposal then was that the concept of love would receive filling if we interpreted it conjointly in terms of

THE ETHICAL DIMENSION

the following: concern for others which primarily reflected our ultimate concern, for God, and his undying concern for us mediated through Jesus Christ; commitment to the others so that our concern did not stand on the side-lines, pitying them and wringing its hands, but committing itself wholly to them in succour in their need; community, the fellowship with others which is a stable, lasting, and demanding thing, not allowing us to withdraw when we become bored, or tired, or drawn away by some new fad or fancy, the whole process of being bound up together with our fellows in the bundle of life; involvement, which emphasises that in the loving communion we are ourselves involved, not holding back or protecting ourselves or our emotions, but exposing ourselves to the possibility of suffering or sacrifice; and finally identification, which seems to gather up all else both as the condition of their happening at all, and as the consummation of them. If I were now to re-trace that map of the world of love, I would give a place of importance to a concept which I mentioned then, but did not emphasise, namely, imagination. I see it as important, both in relation to the structure of love, and as the key to how identification works. First, those features of imagination which are particularly relevant to the analysis of the structure of love are: a heightened perceptivity towards other persons which will discern in them qualities concealed from casual observers, as well as difficulties and sufferings, that are afflicting them, even to the attractiveness that underlies the superficial unattractiveness. True perceptivity of the kind associated with imagination has an anticipatory character, being aware not just of the present but of the sequel to the present, with the dangers and the hurt as well as the happiness that it may contain. Much of the unlovingness that we put into our relations with one another would be removed if we were imaginatively to anticipate the hurt that it is likely to cause. Several of our secular philosophers had something to say about the influence of the images that we conjure up, upon our behaviour, and we shall be returning to it. Meantime, however, the other side of the coin may be noted, that imaginative anticipation generated by perceptivity will impart to love a quality of stability and permanence which lifts it above the level of instinctual emotional reaction. Here, you could say, lies one of the differentiae between what we call a casual relationship and one of true love, that the latter possesses as a distinctive feature a capacity for imaginative sensitivity towards the future of the relationship, seeking to ensure that the quality of that future is not marred by thoughtless or precipitate action in the present.

The second aspect of love to which imagination is particularly relevant is the empathy or sympathy which it involves. The comprehensive term for this aspect of love which we tend now to employ is 'identification' already mentioned. Identification is a process of self-exteriorisation in which the agent projects himself into the situation or condition of another person. This process entails emotional participation in his feelings, intellectual understanding of the situation or condition, and the volition or conational attitude, to be prepared to act in a manner appropriate to the feelings and in the light of the understanding of the condition or the situation. In the process, the agent does not become the other person; he is not absorbed into the situation, nor does he acquire the condition. To identify with a drug-addict or an alcoholic does not imply becoming either of these. This reminder is most important when we become increasingly aware of the range and comprehensiveness of the love which Christ requires of us — the love which loves him only in loving those who are hungry, thirsty, naked, in prison, sick and alien. The identification so essential to love, in thus not becoming that with which it identifies, does not require the agent to reproduce in himself the precise feelings of those with whom he identifies. In fact, the agent in the process of identifying may extend the situation into which he is self-projected, feeling more deeply about it than does the person involved (as is the case in Christ's repentance over our sins); understanding more clearly the truth of the situation and what has caused it; and moved to act more effectively perhaps on behalf of the person affected. This self-exteriorisation in some ways resembles Christ's ingression into sinful human nature, stated by St Paul (II Cor 5.21), 'For God has made him to be sin for us who knew no sin'. Identification thus construed goes much farther than either sympathy or empathy, particularly insofar as it does greater justice to the cognitive, intellectual and conational elements of

THE ETHICAL DIMENSION 75

love, the elements which give it head and will as well as heart. But the means whereby such identification is effected is imagination, which, as we have been observing, perceptively places itself in the other's shoes, understands his feelings and cares enough to take remedial or reassuring action. It is imagination which enables the self-projection into the situation or condition, which in itself may be totally alien to the person observing, alien in the way in which the mean-ness of Zacchaeus would have been to Jesus; yet that alienness is in itself no ultimate barrier to the love which has imaginatively entered the situation and which cares profoundly for the condition.

But this self-exteriorisation which takes place in identification does not happen in any automatic way; on the contrary, it is the outcome of the openness which imagination engenders, a heightened dimension of sensitivity to the needs, the sufferings, the hopes and the potentiality of the other person. The enclosed person, the shut-you-out recluse, who is turned inward is not going to be aware that there are others around with heart-breaking needs. It is significant that J. V. Taylor (*The Go-between God*, p.165) said of a similar situation, 'It is at this point that the Holy Spirit may enable a breakthrough — not, as has so often been thought, by a gift of superhuman power, but a gift of awareness ... He opens one's eyes to see the situation differently'. I think that I would want to say that he imparts to us the gift of imagination which enables us to be totally open and receptive both to what is going on around us and also to what the sequel to that present situation might be. We can here recall what we said earlier about the Holy Spirit being the imagination of God let loose in the world, and about the image of God in man appearing in the form of man's imagination, for imagination as creativity is of the essence of God's being. Recalling that emphasis, we see that such openness as we have associated with imagination is the expression of the image of God in us, we sharing in his sensitivity to the needs of the world and of our fellow-beings.

So when we construe love in these terms, we begin to see that it is wrong to speak about the tension between the absolute principle and the sordidness of the practical situation which reduces the ideal to an irrelevancy. The error consists in thinking of love as some kind of universal which is so perfect and celestial that it finds the obduracy and the evil of the human situation totally resistant to any possibility of its particularisation or embodiment in its midst. It is not a question of the particularisation of an ideal in an imperfect medium, but of the way in which one person or group or community may show to others the love which they themselves in all their sin and imperfection have received from God through Jesus Christ. The applicability of that love to that human situation depends under God not upon the intractability of the medium to which it is directed, but upon the openness of those people to whom God has already shown his love, to others in need. The introduction of the Platonic concept of the Ideal and its particularisation into the interpretation of the relation of love between persons has led to the misrepresentation of how love informs human relationships. It informs them in the way we have been describing, through imagination creating openness, empathy and identification, and when love so expressed becomes effective, we have passed beyond doubt about its relevance or its practicability. In short, it is love itself operating imaginatively and with openness, which resolves this particular ethical tension.

Thus encouraged, we may look at another of our ethical tensions to see whether the concept of imagination can be of assistance to us in dealing with it, the tension, namely, between persons and principles, reflected somewhat in the tension also between principles and situational idiosyncrasies. Perhaps we might cut through some of the tangle if we observed that these tensions imply almost the reification of principles and situations, and consequently also of persons, so that you have in the end the problem of discovering the external relations which obtain among these three entities or groups of entities. Thus stated, the problem reveals just how artificial and contrived it is. What in reality we do have are persons-in-relation, and absolutely no other entities of comparable order or stature; and what have been construed as further entities must be re-described as the ways in which these persons are or should be related to one another. Therefore, when imagination controls such relationships, what it will be looking for will be the ways in which the persons in

THE ETHICAL DIMENSION

the situations can be related to one another according to God's prior love for both of them. There will be no higher entity existing in some higher sphere, some ideal laid up in heaven, like a Platonic Form, for the sake of which the person will now have to be sacrificed. Sacrifice there may now well have to be, but it will be seen to be part of the relationship between the persons as they stand together in the context of God's love and God's will. Nor will there be any question of the sacrifice of the principles for the sake of the persons. If there does occur a violation of what is accepted as a high ethical norm, it will be because, for some quite exceptional reason, the relationship between the persons again in the context of God's prior love for them requires it; and such violation will occur far less often than any permissive ethic, or some facile forms of situation ethics, seem to expect. In fact, such violations as such ethical systems are prepared to contemplate could well be avoided with the aid of predictive imaginativeness, which can anticipate the loss of respect, the souring of the atmosphere, together with the sheer unhappiness, which will before long begin to corrupt the relationship. So, in keeping firmly in our minds that it is the function of imagination to be open to the needs of all parties in a situation, to be perceptive of the ways in which God's love can act as the redeeming power in every situation, and to apply that love to the relations of the persons in the situation, we then begin to see that there can not be conflict between persons and principles, or between persons and situations. For these are not co-ordinate, the persons being primary, the situations being persons-in-relation, and the principles being how the persons are to be related. Therefore when what is normally said to be such a conflict arises, imagination will deal with it in the Spirit in the same way as it deals with more straightforward occasions, in terms of caring and perceptiveness, anticipating the quality of the relationship which will emerge from the contemplated action.

Now, lest I should be thought to be denying that there are such entities as moral principles and norms, let me make my position clear in two ways. First, what I have been arguing against is the false dissection, you might even say vivisection, of the moral situation into a number of parts which are then set in antithesis to one another; ethical situations are composed of persons in relation to one another in the context of God's love for both or all of them. Principles have their place there, as the ways in which these persons ought to be related to each other. Secondly, the principles, if we prefer to find another term for them, could be called guidelines, or indicators to the way in which in the past God has expected his people to relate to one another and to the community and, we must add, to the created order around them. As guidelines their authority derives from their origin, and they are operative for at least 99.99% of the time; they constitute the *prima facie* case for relating to our neighbours in a certain way which they define. The very infrequency of the remaining .01% will in itself make it noteworthy and deserving of special and perceptive attention. Here once again emerges the role of imagination, to discern the niceties of the special case and why it should constitute an exception, just as it had in the 99.99% adopted the validity and the applicability of the guide-lines appropriate to them.

A point of which very little is made in the analysis of such guide-lines is that in Christian morality they do not always consist of universal principles and injunctions. They may quite frequently be parables, with the final line to them such as we have in the case of the parable of the Good Samaritan, namely, 'Go, and do thou likewise' (Lk 10.37). The original questions which prompted the parable were: 'What shall I do to inherit eternal life?', and 'Who is my neighbour?'. Now if a story like that is to be a guide-line to Christian conduct, it is clear that we are a long way from universals and their correlated particulars, though the applicability concept is extremely pertinent, as we may well ask, 'Does this parable apply to this situation?' Both inductive and deductive reasoning fail us here, for they depend at some point on a universal; whereas in the case we are considering, we have argument from particular to particular, or more precisely, to one quite specific and unique situation, with the persons within it also unique, from a story told in vivid image forms, the very stuff of imagination. The imagination applies the parable not by means of some intermediate process of universalising it in order to get an umbrella under which to establish the particular and give it a false security. That sort of logical security does not exist in Christian ethical reflection, which goes from imaged particular by way of imaginative insight and sensitivity, to a concrete and specific situation, and the decision which it requires. We do not deny that there are universalising tendencies in the parables, as we have already observed, though fewer than we would expect are original. The role of imagination in private Christian devotion, particularly as such devtion includes prayer for guidance in life's daily duties, is itself a fascinating subject. For the present, however, we note that wherever Christian moral action seeks to be guided particularly by the teaching of Jesus as we find it in the New Testament, we shall expect that imagination plays a very important part both in relating the proposed action to its guideline in the appropriate part of Jesus' teaching, and even prior to that, in the selection of the applicable guide-line. There are, however, other concepts in Christian ethics to which imagination is just as central as it is in the case of love. I am thinking of forgiveness and the acceptance which should be its normal concomitant, and without which it would not truly be forgiveness. All forgiveness involves an imaginative self-projection into the place of another person who has wronged us. Popular usage recognises this fact in the saying, 'To know all is to forgive all'. Certainly there can be no forgiveness without some perceptive understanding of what was in the mind and the heart of the offender, even when all that was there was sheer mindlessness or heartlessness. It is, as we saw in the study of love, the role of imagination to effect that empathy and identification. In this case it has to take further steps to ensure that the forgiveness effectively reaches its goal in the heart of the person to be forgiven, so that he accepts his acceptance, to use a well-known phrase of Paul Tillich's. For there are many ways to forgive, some of which leave the impression that we are conveying a favour, we who have ourselves been forgiven to more than seventy times seven. Sometimes we forgive grudgingly, savouring the moral superiority which our position gives us. Sometimes even, we forgive because we feel that it is our duty to forgive. In none of these cases will our forgiveness be accepted by the person we intend to forgive, and the whole process will be stultified. So perhaps it will begin to be evident why I insist upon a place for imagination in forgiveness, for I fear that without it, all will be lost. In a sentence, therefore, imagination is necessary both for the initiation of the forgiving process and for the effective consummation of it in the acceptance of acceptance by the forgiven person.

Now, so far, we have been thinking of the ways in which imagination may assist in enabling us to place ourselves in the shoes of other persons. But it may assist in a no less important ethical process, that of self-understanding. This process involves once again a kind of self-exteriorising, not into another person, but so that the 'I' which is known and the 'I' which knows are the same person, but the same person imaginatively looking at himself or herself as if he or she were another person. This kind of self-exteriorisation is at the heart of a whole nexus of Christian activities: penitence, shame, contrition and guilt. We have to look at our past as if it belonged to someone else, so that we can with at least a degree of objectivity and without the customary rationalisations, pass the sort of judgment on it which we are all too ready to pass on the actions of others. Ordinary thinking does justice to part of imagination in such activity when it refers to the 'image' that we have of ourselves, and the 'image' which we try so hard to project. It is imagination which handles images, and it is not at all beyond falsifying the evidence, as the Old Testament (Authorised Version) so openly recognised. But when the imagination operates authentically, then what begins as self-knowledge and honesty with self becomes the therapeutic process of shame, penitence, contrition and guilt, and from the side of God, forgiveness, acceptance by him and renewal. Here the imagination seeks to create through openness to self, openness to God.

Perhaps it is just there that the imagination may have something to contribute to what may well be regarded as the most serious of all the tensions that we mentioned, namely, that between the good that I would and yet do not, and the evil I would not and yet do. For openness to self and honesty about one's weakness is the point at which God can begin that penetration of our wills which strengthens us to do the good we would, which is what he requires of us. That will not be all that is necessary, but it will be the start; and it is also

THE ETHICAL DIMENSION

suggestive of the fact that imaginative perceptiveness which keeps us open to ourselves may have the further role of keeping us open towards God from whom the power comes to do that which he asks of us. For self-knowledge will also entail an understanding of the ways in which we endeavour to shut God out as we so often shut out our fellows. So then, the Spirit of God who, as we have said, is the imaginative Spirit making us in God's image, will keep us open to that power which, in forgiving, renews, and, in renewing, invests us with a new authority of mission and service.

There is, besides, a second way in which imagination may help in this tension between our will to do good and the reality of our continued wrongdoing, and it may be explained by a reference to a saying of Jesus, in Mt 5.27f., 'You have heard that it was said, "You shall not commit adultery". But I say to you that everyone that looks at a woman lustfully has already committed adultery with her in his heart'. The traditional interpretation of these verses is that lusting is crypto-adultery, which is none the less reprehensible for being unfulfilled. Another interpretation which is just possible, though I would not press it, is that by lusting we conjure up in the imagination images or fantasies which make the fulfilment of the lust, when opportunity offers, all the more probable. As Hume would have said, we have a 'lively impression' of the self, and in this case, of the self acting in a certain unacceptable way. If such images or fantasies are repeated, they build up a volitional disposition in a certain direction, so that even rather belated attempts to restore the balance on the side of chastity are doomed to failure. Imaging of this sort is to be found over wide areas of our mental and spiritual life; it is at the heart of all ambition, both good and bad; it is an essential feature of any planning of a programme for our personal future, for our church, or for our relationships with others, which require some special attention. Such imaging is to be treated with the greatest of care, and nowhere more than in relation to the tension endemic to our whole existence, the tension between the good we admire and the evil we accomplish despite all our good intentions.

Another way to describe the phenomenon we are examining is to see it as a classical case of Wittgenstein's 'seeing-as'. I see myself *as* doing this or that, *as* being this or that kind of person if I do this or that; and such self-images are devastatingly potent. They are particularly relevant to the consideration of another aspect of all morality, namely, the question of decision-making. Very often under the influence of the determinists who are anxious to demonstrate that there is no freedom in our decision-making, since it is socially, economically or genetically conditioned, we tend to interpret it retrospectively as the resultant of a parallelogram of forces. There used to be a much publicised distinction in moral philosophy between motive and intention, the one being the force from behind which powered an action, the other being some future state or result which, it was hoped, the action would achieve and which attracted the action *a fronte* into fulfilment. It is in the area of the intentionality of an action that imagination has a major part to play, for it is its role to be percipient of the future, to anticipate results, and — a point which we have so far omitted — to invest these results with emotional cladding. Therefore decision is not simply a semi-mathematical computation of the forces that push in the direction of one course of action, and a comparison of that total with a similar figure in favour of another course of action. There may be a small element of such assessment of what we may call for this purpose the motivation of the two courses. But much more important is the intentionality of both, the image of ourselves as doing the one action, or as doing the other, the image of the kind of person we would be in either case, and the emotional toning that accompanies the images in the two cases. Operating all the time, one would hope, in decision-making which takes this form, would be the imaginative self-critical mechanism which some psychiatrists have called the super-ego, and so creates self-knowledge. Construed in such terms, decision-making is wrongly regarded as the summation of your character at the moment of deciding, and consequently of acting; it is, rather, creative of your character at these moments, for by it you are dynamically forming the shape of the person that you are going to be.

It is in the context of such a construction of the process of decision-making that I would like to approach the tensions associated with freedom and determinism. We have partly

THE ETHICAL DIMENSION

indicated an answer to this problem, for it is at the point of decision-making that the controls and constrictions of determinism will be considered to operate. Now I would not pretend for a minute that many of the things which determinists say are not without considerable truth. Social, economic, educational and genetical forces do operate not only with motive force when we set about making decisions; they also succeed in limiting the range of self-images that we may conjure up, and in seriously affecting the quality of the emotional toning that accompanies them. But just because these are images of what is not yet, lying in the future, they can not be contained within motivation; they lie in intentionality, and it is there freedom resides. I find the contribution of Sartre to this subject most useful. Like all the existentialists, he is deeply aware of the parameters within which human beings exist and act; but he locates in the imagination the power of the human being to transcend these conditions and conditionings, and to create a world which in spite of its rejection of much that is in the world remains nevertheless in continuity with it, and has not disappeared into a fantasy world. Such imaginative activity of the mind is of the essence of consciousness, which according to Sartre could not exist without it. Indeed, without it, we would have to say that the determinists were correct. But existing, it remains untouchable by the determinists. In the hands of the Christian apologist, the imagination may be presented as an enormous power for good, or alternatively as a power for evil. Therefore, freedom expressed in terms of human imagination is a two-edged power, as in fact so many of the gifts are which God has placed at our disposal; we may use them or we may abuse them. That same freedom has to be conserved in any theories we entertain concerning the transmission of values within an educational system; but it will be freedom to be exercised, not neutrally across a pluralistic range of options, but creatively within a system of imagery presented with values and with affective toning. No effective educator has ever achieved the aims of education in his field without a form of presentation which reflects such values and evinces such toning, and so imparts the criteria by which they are to be assessed.

Imagination as a Devotional Principle

It requires very little extension of what we have been saying to describe the way in which imagination is related to our devotional life. In fact several points come readily to mind. For example, insofar as self-examination is a customary part of devotional exercises, it will only be effectively sustained through the employment of self-images, which through a process of self-exteriorisation enable us to achieve self-understanding. That in turn leads to the possibility of penitence, contrition and finally confession, with its promise of the forgiveness of God. Therefore, right at the start we see that the pilgrimage of the soul from the sorrow of self-knowledge through to the appropriation of divine forgiveness is a progression which is at least facilitated by imagination. Expressed in the highest terms, the progression is initiated by the Holy Spirit who is the imagination of God, and sustained by him through his image in us which is our imagination. The imagination has been observed also in the decision-making process, which is for many Christians carried out in the context and atmosphere of Christian devotion. The imagination serves up the images of the self acting in this way or that, and images also of the kind of self which will emerge if different courses of action are followed. Out of such images, and through the comparison of them with one another, rather than as a result of pressures from behind, does decision eventuate.

Let us now detail further aspects of the devotional life to which imagination has a contribution to make. First, we read the Bible for the simple and obvious purpose of understanding who God is, and of what character, what he has done and continues to do, how he has achieved his purposes in and through the vicissitudes of human history, and what his present offers to, and demands of, us are. These subjects are presented to us in the Bible not as ready-made doctrines, but as stories about this event and person, and that. To read such a book, or rather series of books, as the Bible, demands that we should endeavour so to project ourselves empathetically into the material before us that we re-enact it, as Collingwood would have said, in our own thought. This fact becomes all the more important when we remember that what the Bible has to teach about God is given to us not in the abstractions of conceptualised propositions, but in the concreteness of stories. Even such intellectually difficult and complex notions as that of God the Creator come to us first in the story of the earth being 'without form and void', of the Spirit 'brooding upon the face of the waters', and so on. Such story forms are of the essence of imaginative thinking, and have to be constantly returned to, even when we have employed other, conceptualised, forms of thought. Where we would theologically want to speak of the transcendence of God, the Bible speaks of God 'high and lifted up, his train filling the temple'. In the New Testament, the imaginative approach comes into its own, for there we are dealing with records which have an historical foundation, and for their understanding require an empathetic self-projection into their inner character. Such understanding is both the source and the justification of the sophistication of our Christological analyses.

Secondly, in Christian devotion the Bible is used as the means of worship and the way in which the worshipper gains access to the promises, the comfort and strength of God. I am thinking of the Psalms with their rich expressions of the glory and the majesty of God, of his infinitely tender care and compassion, all given to us in a series of images. So when in, for example, the Benedictine offices we sing praise to God in the Psalms, we do so by re-enacting the thought and emotion of the Psalmist, and by re-living the past expression of praise which comes to us across two and a half thousand years. So, imaginatively also we place ourselves where the men of the Bible stood when they received the promises of God, the assurance that God's help would be forthcoming, that he would be their keeper on their right hand and on their left, the one who would guard their going out and their coming in, and that he would abundantly pardon. When we come to the New Testament, then imaginatively we place ourselves in devotion in the situations where Christ was, becoming what Kierkegaard called 'contemporary disciples', so that what was said in Nazareth, in Jerusalem, at Bethany, and in Gethsemane, is said to us now. It is through such sustained imaginative contemporaneity that the Bible has remained the supreme medium of Christian devotion.

But there has also been a long tradition of seeking the will of God for a specific situation through 'searching the Scriptures'. Such an activity, as we have seen from the ethical part of our discussion, is essentially an activity of the imagination, and it is so in a double way. Previously we noted this as a special form of ethical thinking. It does not proceed from universal principles by way of an Aristotelian practical syllogism, but goes rather from the understanding of a parable or a biblical situation which was quite specific and non-general in character, to a further quite specific course of action. Now we are adding that such an application of a New Testament passage or incident to a here-and-now ethical situation is conditioned by our being able imaginatively to enter into the intention of the original situation, and equally imaginatively to make the transition to a world and a time remotely distant from, and in so many ways alien to, the original.

It is perhaps in relation to prayer that Protestant theology has to make its greatest reassessment of the part played by imagination and images. We spoke earlier of the mistrust of images which is endemic to this denomination, extending from suspicion of images and ikons in church through to the wide substitution of concepts for images in thought about God. Such reassessment will not be effected overnight, but we can begin to indicate the direction in which the first steps might be taken. For example, we should not find it difficult to carry over into devotional life the image-forms in which we saw the Bible speaks and thinks of God, instead of the conceptual structures which theology imposes upon the biblical language and stories. But the imaging need not be only linguistic; it may also be visual, and we shall then recover for one denomination what several others have found to be singularly helpful, the use of images and pictures for the concentration of thought rather than for its distraction, and for the filling out of our prayers rather than for their deprivation. Each of the individual modes of prayer is immensely benefited when it is 'ikonised', transferred into image-forms; adoration, thanksgiving, confession, supplication, intercession — all become more specific and precise the moment we cease to regard prayer as the extension of our theology, and integrate it into the normal currency of daily thought and work. Successful meditation of an articulate type can be built up, through the creative use of symbol or a system of symbols, through the sustained contemplation of a picture, or a sculpture, or a cross. I would therefore resist somewhat any easy surrender to the influence of oriental forms of meditation, which insist upon abstraction from images and also concepts, and a withdrawal into emptiness of mind and spirit, which is the beginning of spiritual communion with ultimate being.

CHAPTER 5

IMAGINATION
THE PHILOSOPHICAL DIMENSION

There can be very little doubt that one of the major causes of the failure of imagination to have found a place within Christian theology, and philosophy of religion and epistemology, has been the extremely unsatisfactory treatment which this mental activity has received within philosophy itself. It has appeared and re-appeared in philosophy ever since Plato gave it the lowest place in his fourfold division of the line of knowledge, in *The Republic*, Book VII, but it has never achieved that consistency of analysis given to the subject of sense-perception, or to rational knowledge on the other hand. It moved about in a no-man's land between the two, so that Rodney Needham (*Belief, Language and Experience*, Blackwell, Oxford, 1972, p.134) could with all justification say, 'This word fulfils a variety of odd-jobs, severally interpretable in different contexts'. I propose to do two things in this attempt to find what the philosophers, or at least, a selection of them are saying about imagination that might be of help to us in our theological enquiry: first, to show just how varied these odd-job functions are, according to the context; and secondly, to argue that there is nevertheless within the history of philosophy, evidence of a stable and continuing role which imagination can be recognised as fulfilling, particularly in the field of epistemology. First of all, *the odd-job functions*:

One of the most fascinating, and far from exhausted, interests of many writers on imagination, concentrates on the *equation of imagination with imaging*. It is an equation which we have already observed in George MacDonald. In *Mental Images*, pp.2f., Richardson defines such entities as referring to (i) all those quasi-sensory or quasi-perceptual experiences of which (ii) we are self-consciously aware, and which (iii) exist for us in the absence of stimulus-conditions, that are known to produce their sensory or perceptual counterparts, and which (iv) may be expected to have different consequences from such sensory or perceptual counterparts. The points that are thus made are that images possess a peculiarly sensory quality, are non-general and quite specific, do bear a certain relationship to the sense-perceived world of ordinary experience, but are not, as regards content, externally stimulated in the ordinary way. At this point a fiercely contested controversy has developed. On the one hand, Ryle (here standing very close to Jean-Paul Sartre, as we shall later see) denies hotly that there are any such things as 'mental pictures' (*The Concept of Mind*, p.241). The activities of picturing, visualising and 'seeing' are all proper concepts, but their use does not entail the existence of pictures which we contemplate, as if they were photos 'in the mind's eye' (p.234). Interestingly, Ryle agrees that 'imagery occurs, but images are not seen', just as tunes 'running in the head' are not heard. Sartre makes an additional interesting comment. While like Ryle, not denying that images exist, he affirms that to imagine (or image) an object is to be related to it in a special way, in an act in which an absent object or person is envisioned as present. Images, therefore, are the way objects that are absent have of being present. On the other hand, there are two major writers in the field who represent a position diametrically opposed to what we have just heard. Alastair Hannay, (*Mental Images: A Defence*, Allen and Unwin, London, 1971) says that 'no good argument for getting rid of mental images is forthcoming' (p.23), and that 'there is no linguistic strain in calling imaging a kind of seeing' (p.149); while Mary Warnock, now Dame Mary Warnock, (*Imagination*, Oxford, London, 1976) contended that 'it makes no sense to say that "there are no such things as images"' (p.182). The truth is that there is at present in the philosophy of mind a great deal of confusion about what images are, what their intentionality is, what their ontological status is, and so on. The application of any substantial content of the philosophy of images as we have here been discussing them to theology is not likely to be of enormous help, beyond two points which I would like to make. First, while there seems to be a major disagreement between Sartre and Ryle on the one hand, and Hannay and Warnock on the other, they do not disagree about the fact that there are *images*, that they do not have the qualities of percepts, and that they seem to be correlated to a form of apprehension which is 'seeing' as distinct from seeing *simpliciter*. This type of 'seeing' (with inverted commas) looks rather like a mental activity not dissimilar to faith, which has been sometimes so described. In other words, both faith and imaging could be regarded as ways in which we think about certain objects, or subjects; and it is as valid a form of consciousness as perception, or conception, or inference. Secondly, Sartre's concept of images being the way the absent is present has again overtones of significance for understanding faith's awareness of God, who though not visibly present, and so in that sense absent, through our modes of apprehending him, is present.

The next odd-job function which imagination performs occurs in *the form of imagining-that*. The job here performed is that of mentally constructing a state of affairs, a situation, an occasion, by collocating, inter-relating and/or setting in sequence, particular objects, occurrences, or persons. Clearly we have here moved into a much more complex odd-job than simply forming images. Imagining-that may involve, and often does, the calling up of images which may be percept-derived, while others of the components of imagining-that constructs are non-sensuous; for example, some relations which figure among such components are non-perceptible. Simply, we might imagine-that there is a unicorn sitting at the top of the Scott monument; in a more complex mental act, we might imagine-that democracy has sufficient moral power to outlast all forms of totalitarianism finally within history. While both democracy and totalitarianism may as concepts have both concomitant images attached to them when we think about them, especially if we have been looking at some caricatures presented in *Punch*, where they have appeared; nevertheless both they and the relation alleged between them can feature in an imagine-that situation without recourse to images. Further, 'imagine-that' has to be compared with, and distinguished from 'suppose-that'. According to E. J. Furlong, *Imagination* (Allen and Unwin, London, 1961, p.27) there are two cases where 'imagine-that' could be used instead of 'suppose-that': namely, first, the case of plain supposal, as in 'Peter, suppose that you are an ice-berg'; and secondly, if we are devising some kind of map, we could well say, 'Let us suppose that this point represents a Church'. In the second of these cases, Furlong is right to add that 'suppose-that' equals 'postulate-that', as we might do in an Euclidean theorem, where it is clearly not replaceable by 'imagine-that'. He offers another example which I find interesting, namely, 'Let us suppose that we have a space of ten dimensions', where he argues that it must be rejected as a case where 'imagine-that' might be substituted for 'suppose-that'; the reason he gives is that the supposal conceived of is something which can not be pictured, or visualised, or sensuously imagined. Such a reason for the rejection of this case is not adequate, because it is wrong to confine 'imagining-that' exclusively to cases where the subject is sensuously imageable, as we have just been arguing in the example about democracy and totalitarianism. This point is certainly worth emphasising at this stage and in this context, because so much of the imagining which appears in the field of religion and theology will inevitably have to do with non-sensuous material. Another theological connection with 'imagine-that' can not be missed, for there is a sense in which it is the bracket, or the quotation marks, which preface the parables of Jesus. If so, the connection tells us something about the content of the parables, namely, that we shall expect them to contain sensuous and non-sensuous material; and, more importantly perhaps, about the frame of mind in which Jesus in the first place approached the parables, and in which he expects us, following his example, also to entertain them. They are the products of imaginative creativity, and are not to be handled with heavy literality. A third odd-job performed by the imagination appears in *the form of 'imagine-how'*. Edward Casey, (*Imagining: a Phenomenonological Study*, Indiana University Press, Bloomington, 1976, pp.44ff.) in a very interesting comparison of 'imagining-how' and 'imagining-that', says that there are two similarities between them, within which similarities occur a striking dissimilarity in each case, thus. In the first place, 'imagine-how' like 'imagine-that' is usually the introduction to some state of affairs or set of circumstances:

'Imagine how you might escape from a sealed barrel under fifty feet of water'; 'Imagine how you might survive on one ounce of rice per day'. It is equivalent to imagining what it would be like to be or live or act under the circumstances described; and it presupposes an 'imagine-that' to provide the setting of these circumstances, for example, 'Imagine that you have been put in a sealed barrel, and that this barrel has next been submerged in water. Then imagine how you would escape'. The dissimilarity lies in the fact that in the case of 'imagine-how' the imaginer is not a mere witness (as he is in 'imagine-that'); he is a participant in the experience. Casey adds that the imaginer *identifies* with the experience, and develops a sense of personal agency. In the second place, 'imagine-how' like 'imagine-that' can occur either sensuously or non-sensuously; but whereas 'imagine-that' is the preface to a static set of circumstances, 'imagine-how' involves the imaginer in a dynamic state of affairs 'with rhythm and movement' (*id.*, p.46) and once again he identifies with the person instead of being an onlooker. Three comments may be made on what Casey has said here. First, to tidy up the record: he does not make anything of another distinction between 'imagine-how' and 'imagine-that', namely, that the latter has always to be followed by a noun clause, while the former may be followed either by a noun clause, as in the examples given above, or by an infinitive, as in the sentence, 'Imagine how to do certain things, to speak or behave in a certain way'. Secondly, perhaps Casey should have noted that *grammatically* the clause introduced by the 'how' of 'imagine-how' is itself a noun clause of which the paradigm case is introduced by the word 'that'. Thirdly, the theological connection is not by any means remote, especially if we pick up Casey's use of the notion of 'identification', which we have already examined in the study of the ethical dimension of imagination (*vide supra*, pp.65–87). The intention of the parables of Jesus is not merely to occasion us to 'imagine-that', to remain witnesses of an imaginary scene, but rather to bring us to the point of imagining-how it feels to be left for dead on the roadside between Jerusalem and Jericho, with help failing to come from the sources that could have been expected to give it; or of imagining how we would react if the bridegroom for whom we had been anxiously waiting greatly delayed his arrival, or we had found a pearl of great price. Jesus' injunction to 'Go and do likewise' is not an addendum to the story; it is the making explicit of the purpose for which the parable was told in the first place and which was embodied within its whole structure. The parables, in other words, require us not just to 'imagine-that', but also to 'imagine-how'.

Before leaving this rather semantic analysis of how the word 'imagine' is used by us, with its message that these senses have all percolated into theological and religious employment, I would like to draw together a cluster of odd-jobs that imagination does, along the lines of that previous analysis. I can think of 'imagine-where' as in, for example, 'Can you imagine-where he might have hidden the treasure?' and 'I certainly can imagine-where in *Phantastes*, C. S. Lewis found MacDonald's work so inspiring'; and 'imagine-when' as in, for example, 'I can not imagine-when he would have come to his senses, if bankruptcy had not overtaken him'; and 'imagine-whether', as in, for example, 'Can you imagine whether it is meanness or poverty that keeps him from entertaining his friends?'; and 'imagine-why', as in, for example, 'I have tried every way to imagine-why he did not complete that book'; and 'imagine-which', as in, for example, 'Study this script very carefully, and see if you can imagine-which of the characters is based upon real-life persons in the College'; and, finally, 'imagine-what' as in, for example, 'Imagine-what an opportunity exists in Australia with so much sunshine, to grow so many exotic flowers', and 'Imagine-what it feels like to bathe in coral seas'. My comments on these cases will be brief. They all introduce states of affairs, circumstances, events, or persons in relation to one another. But not only do they all presuppose an 'imagine-that' previous state of affairs, they also require some action from the imaginer in relation to that state of affairs. Accordingly, they resemble somewhat the 'imagine-how' case insofar as they call for involvement, or identification, or imply them. There is perhaps one exception to that statement, namely, the example, 'Imagine-what it feels like to bathe in coral seas', which looks singularly like a straight 'imagine-that' case; though if we were to substitute 'Imagine-how it feels to bathe in coral seas' for 'Imagine what it is like

THE PHILOSOPHICAL DIMENSION 95

...' the exception disappears. Also, as was the case with 'imagine-how', the clauses introduced by 'where', 'when', 'whether', 'why', 'which' and 'what' are all noun clauses of which the paradigm is introduced by 'that', but they all share the participation implication already seen to be the norm with 'imagine-how'. It could be said, of course, that the word 'imagine' in a number of these examples is being used in a weak sense, and is really equivalent to 'guess', 'estimate', 'judge', or 'suppose', and therefore we should not be leaning too heavily upon the nuances of meaning which it seems to carry. Finally, these illustrations serve to show the range of the circumstances in to which imagination may penetrate; and from the theological point of view should prevent us from isolating it to one single type of verbal construction or psychological experience or activity.

Our second main theme in this chapter is *the role of imagination in epistemology*.

The previous discussion of the so-called odd-jobs which imagination performs, illustrated by the variety of situations in which it operates and how it operates, has taken us to the centre of the epistemological role which imagination plays, what might be regarded as its bread-and-butter tasks. It is beyond the scope of our present purpose to provide at this point a comprehensive history of the treatment by philosophers of the notion of imagination across the ages. I shall, on the contrary make a quite personal selection of philosophers, not because I think that they are the most important for several of the more important have been discussed in other works (see particularly the excellent *Imagination* by Dame Mary Warnock, previously mentioned), but because they open up fresh and different vistas of relevance to our own purposes.

Plato

A start will be made from *Plato*, where so much else in philosophy began, and who was destined to have such a great influence on Christian theology for centuries later. We shall examine Plato's definition of the place of imagination in the following four areas: in the first stage of knowledge; in mathematics; in the arts; and finally, in ethics. *The Republic*, to which we turn first of all, is a complex work, combining ideas on the nature of justice understood in social as well as private terms, on the nature of the education of the young, on the nature of ultimate reality and the so-called Form of the Good. We shall be concentrating to begin with on Books VI and VII of this work, where he sets out the spectrum of different forms of knowledge and their objects, by means of what he calls 'the divided line'. He first gives the subject a pictorial representation by means of the now famous parable of the prisoners in the cave. They are confined to a subterranean cave, unable to rise or turn round, condemned to face the wall at the back of the cave, on to which are projected shadows thrown by a distant and to them invisible fire in the upper world. A parapet has been set across the mouth of the cave, and behind it pass men carrying figures of animals and imitating their cries. The parapet conceals the carriers but allows the shadows to appear on the wall of the cave; and the shadows become for the cave-dwellers the only existing world of things. Such order and connection as the shadows have derive from the outside world, and on the basis of such order and connection, the prisoners learn to predict successions of shadows. Eventually, one of the prisoners is freed — Plato nowhere explains how — turned around and taken out into the world. At first, he is confused when seeing the originals of his shadows in the cave; the glare of the sun's rays will blind him; and so he will look at the reflections of objects in water. Finally, he will be able to look at the sun itself, and come to infer that it is the source of the seasons and years, and controls everything that is visible, and that it is the author of all that he and his fellow-prisoners used to see on the walls of the cave. Plato rounds off the myth with the reminder that it is the duty of the freed prisoner to return to liberate his unfortunate colleagues, though he can expect no grateful response from them, but only ridicule and the threat of death.

The theme stated by Plato in such vivid pictorial terms is examined more closely in his presentation of what is known as 'the divided line', and there has been a long-running controversy concerning how the Line and the Cave are related or

THE PHILOSOPHICAL DIMENSION

rather correlated one to the other. The detail of that controversy is not our concern, and its minutiae are not relevant to our purpose; we are interested rather in the main outline of Plato's subject. The line may be illustrated as follows:

	The Sun		**The Good**	
	The visible		The intelligible	
	— doxa —		— episteme —	
	eikasia: imagination	*pistis:* faith, conviction	*dianoia:* understanding	*noesis:* reason
The Line	A	B	C	D
	images: *eikones* reflections: *phantasmata* shadows: *skiai*	things made: *poiemata*	mathematical concepts: *mathematike*	Forms: *morphai*

Now for the exposition: The two worlds of the visible and the invisible are represented by a line (*op. cit.* 509D) divided into two unequal portions, differing in 'their comparative clearness and obscurity'. The line is further divided into similarly unequal portions so the A:B::C:D. The objects in A (*eikasia*-imagination) comprise *eikones*-images, sub-divided into *skiai*-shadows, mentioned as we have seen, in the myth of the cave, and *phantasmata*-reflections or phantasms, such as are seen reflected in water, and images of objects in the material world. Objects in B are 'ordinary living creatures, everything that grows or is made', while *pistis*-faith or conviction is the mental set appropriate to these objects. Section C of the line consists of the so-called mathematicals, *mathematike*, apprehended in the process of understanding, *dianoia*. Section D consists of the hierarchy of the Forms within the Form of the Good as the supreme principle of being and knowledge, *noesis*-reason here being the medium of comprehension. M. W. Bundy says (*The Theory of Imagination in Classical and Medieval Thought*, University of Illinois Press, Urbana, 1927, p.207) that 'for each of the two realms of Becoming and Being, Plato has a theory of imagination', that is, for A and C. Taking A first, we might say with Paton that *eikasia* is 'the first and intuitive vision of the real'. Its object is 'what appears', *to phainomenon*. For it the distinction between the real and the unreal has no force or meaning, for it is cognition of the only kind available at this stage. It does not affirm or deny or lay claim to objective truth. If, therefore, we call its objects 'images' or 'reflections', we must not be thought to imply that *eikasia* mistakes the image or the reflection for the reality which it images or reflects. For it they constitute reality, though the term would not occur in *eikasia*. It has been validly pointed out that there is a long-standing tradition in the history of philosophy which denies that there is any other way of knowing; empiricists from Protagoras to Hume have identified something very like Plato's *eikasia* with the whole of knowledge and its objects with the whole of reality. We observed that Plato conceives of imagination as operating at two points on the line — as imagination of material objects (A) and, since C bears the same relation to D as A bears to B, as imagination operating in relation to the Forms, in the medium of the so-called 'mathematicals'. Here we stumble upon a veritable mine-field of controversy, particularly over what exactly the mathematicals are. The illustrations which Plato offers, as is the case with so many of our own illustrations, are the central source of disagreement. With a proper modesty let us try to tease out what Plato is saying. His illustrations of mathematicals are: 'the odd and the even', 'three kinds of angle' and 'figures', all of which are described as examples of the 'hypotheses' which the mathematician uses as his starting-point, but does not at this stage examine. It is *noesis* which at the next stage of knowledge, D on the line, goes beyond the hypotheses to the *anhypotheton*, the 'unhypothetised' or the unconditional, so close in identity to the Form of the Good. It is how, according to Plato, the mathematician thinks in dealing with these hypotheses that connects with our interest. Thus, though the geometer uses a diagram for the purpose of demonstrating his conclusions, the object about which he is discoursing is not that particular line or figure, not this square ABCD or this triangle ABC or diagonal AC. He is arguing about the square or triangle or diagonal *as such*. His statements refer to *any* such figure or line. The geometers 'use the visible squares and figures, and make their arguments about them, though they are not thinking about them, but about those things of which

THE PHILOSOPHICAL DIMENSION 99

the visible are images'. Two comments may be made at this point. One is that Plato is here saying very much what Euclid was to say much more formally within his system, when in using specific figures he was careful to enunciate that he regarded triangle ABC as standing for *any* triangle, and not to deduce from the properties of this triangle conclusions which might be specific to the particularity of this triangle. In face of the sheer impossibility of enumerating all the cases concerned, the only logical expedient is to treat the case before us as *any* example of the entity concerned. My second comment is that Plato is here associating imagination with what is a most characteristic form of human thinking. Neither is it about universals as such and *per se*, nor does it endeavour to establish relations between universals as class-groups. It uses the particular to prove some theorem concerning the universal. Further, it does not concentrate upon images, and operate by association or contiguity. On the other hand, it does not operate with abstract, non-image concepts; it employs the images in order to think about the universal, or it thinks about the universal through the images.

We can not leave *The Republic* without taking note of what Plato has to say about poetry and the arts in Book X. He is here returning to a theme which he had already mentioned in Book III, where he described poets as 'imitators', who present the actions and thoughts of many imaginary persons, and where he deprecated their theatricality. He did so for two reasons, the first of which was that their life-like presentation of evil is unavoidable in a dramatic picture of life, and it is dangerous to expose good and bad characters with equal realism. His second criticism had been that young people are thereby encouraged to approach life in a theatrical, and somewhat artificial, way. It is not the poet's or dramatist's work as such which is being condemned, but its pedagogic consequences. Now, however, his case becomes much more sophisticated as he proceeds to provide the metaphysical foundation for the earlier strictures, and he begins by affirming the existence of three things: first, the Form, the Idea of, say, a couch, which Form or Idea exists *en te physei*, in the world of ultimate reality which has been created by God, the *phytourgos* (the Creator); individual couches made by the upholsterer, each embodying the Form in a particular occasion, but none a perfect embodiment; and thirdly, the particular representations of such an article in painting, each a perspective of a solid piece of furniture. The upholsterer looks to the Form, but does not succeed in reproducing it perfectly; the painter looks to the upholsterer's model and not to the Form, so his picture is a copy of a copy. He is three removes from reality. The writer of a tragedy is three removes from the King who is God. Art lives on illusions. We should observe that Plato is not answering the question: should poetry and art necessarily deal in illusion? Rather he takes poetry and art as he finds them, and criticises them accordingly. In the paragraphs following, he explores the consequences of watching plays by comic poets, the chief danger being that we should ourselves become buffoons — a solemn warning surely to all fans of Laurel and Hardy, or the Two Ronnies. Plato is here recording the implications of art and the way it uses imagination from the stand-point of education, which is strictly in line with the general educational direction of the dialogue as a whole.

It is very interesting that Plato comes later, in *The Sophist* and *The Laws*, to modify this low, dismissive assessment of the arts, but he does so by modifying his account of the phenomenal world, to allow it a reality which the philosopher must acknowledge and explain. The naturalistic artist is no longer to be banished from the Ideal State. He will be judged not by ideal standards, but by the standards which his own practice suggests. The artist who imitates material reality is to be judged by his knowledge of what he imitates, and by his attitude towards his subject. The imitator, ignorant of what he imitates, is to be condemned. So too, the creator of ideas, not of things or impressions, is the kind of creator that Plato himself was, and he would be essentially an imaginative man. In *Philebus* (40D Jowett's text, 4.611), a reference with a close connection with the interests of *The Republic* occurs, namely, the association of imagination with morality. He has just been describing rather graphically the way in which the memory works, rather as a painter who draws images, in the souls, of things which have been perceived, such images being accompanied by certain opinions about them. The term 'phantasm'

THE PHILOSOPHICAL DIMENSION

is used both of the opinion and the image, and when the relation between the image and the original from which it derives is accurate, then the opinion is true; if not, it is false. But we also have opinions about the future which are accompanied by phantasies, which are expressions of our hopes and fears. They become regulative of our conduct. 'And the phantasies of hope are also pictures in us; a man may often have a vision of a heap of gold, and pleasures ensuing, and in that picture there will always be a likeness of himself mightily rejoicing over his good fortune'. He goes on, 'the good keep before them good images, while the bad have only ludicrous imitations of the true. Their images, the concrete embodiment of their false opinions, lead only to a pleasure about things which neither have nor have ever had, any real existence' (*id.*, 4.612). So the power of phantasy exerts control over conduct — a point that we considered at length when examining the role of imagination and images in an ethical context (*vide supra*, pp.80–83). Self-knowledge will involve awareness of one's phantasies as the form taken by one's hopes, fears and desires. The case has been argued that this conception of the role of imagination was carried forward into Aristotelian and Stoic ethics, eventually to pervade classical and medieval ethics. One form of the influence is to be detected in St Thomas, who derived much of what he had to say about the imagination from Avicenna, and who warned that the intellectual habit of the soul must cut down and repress the presentations of the imagination; otherwise the person will be the less fit to form right judgments. 'The intellect must rule over the imagination' became a principle which was to dominate thought about the imagination until modern times, in art as well as conduct. St Thomas also held that the different actions of the soul are to be assigned to different powers, and so adopted the faculty psychology which was to be so popular for many centuries; and which obliquely also spread the notion that the imagination is one isolable faculty of the mind — a subject which shall come up for later examination.

David Hume

Though Hume is not a philosopher to whom one would readily turn for light on a subject sympathetic to religion or theology, nevertheless his treatment of imagination and the importance which he attaches to it in his analysis of human nature require us to pause to consider another thinker who makes a special contribution to our enquiry. Let Hume speak for himself (*A Treatise of Human Nature*, I.iv, p.225): 'I must distinguish in the imagination betwixt the principles which are permanent, irresistible and universal; such as the customary transition from causes to effects, and from effects to causes: And the principles which are changeable, weak and irregular ... The former are the foundation of all our thoughts and actions, so that upon their removal human nature must immediately perish and go to ruin. The latter are neither unavoidable to mankind, nor necessary, or so much as useful in the conduct of life ...' So he had said earlier (*id.*, I.i, p.13), 'Its effects are everywhere conspicuous; but as to its causes, they are mostly unknown, and must be regarded as original qualities of human nature, which I pretend not to explain'. There are six areas of human thought and action in which Hume defines with emphasis the role of imagination, namely, abstract ideas, the distinct and continued existence of body, the causal relationship, the identity of the self, historical knowledge and sympathy. We shall take them in that order, though our study of them will inevitably be briefer than would be necessary in a more philosophically-oriented study.

We begin with *abstract ideas*. In theology we tend to employ considerably more abstract ideas than the particularistic character of religious situations would appear to justify. Image-thinking, like parabolic thinking is, on the contrary, particular and specific, so that it is not without relevance to see how Hume relates the two. The process of abstract thinking is as follows: We observe a resemblance among several objects, and despite differences in degree of quality and quantity, and whatever else, we apply the same name to all of them. This is the name of the abstract idea, what a later psychology would perhaps call the generic image. So doing, we acquire a custom or habit, whereby the hearing of the name revives the idea and activates the imagination to image it in all its particularity. It is the role of the imagination to deploy the image or images as prompted by the word. The images are held, as it were, *in retentis*, to be summoned up as required. In parenthesis, we might compare the way in which in theology, when a question is asked about, say, the righteousness of God, especially if an answer is expected to take account of the events in the human life and death of Jesus Christ, we would very probably reply in terms of particular situations in the Incarnation, and of images drawn from the records, in which traditionally God has been thought to demonstrate his righteousness.

As regards *the distinct and continuous existence of body*, a simple experiment meets the case. When I look out of my window and see mountains, houses and trees, then look away or shut my eyes, and look back again at them, my impressions of them are the same as before: they are constant. On the other hand, bodies change; the fire which was bright in my fire-place before I went out, has now dropped to a red glow. Changes of this sort which conform to a coherent pattern are also characteristic of external objects. So the imagination is said to have a very peculiar characteristic that 'when set into any train of thinking, (it) is apt to continue, even when its object fails it, and like a galley put in motion by the oars, carries on its course without any new impulse'. So the mind, having received coherent impressions from objects as they have been presented to the senses, continues in the same direction, to make the coherence as complete as possible by supposing the objects also to be continuously existent. We have observed constancy in our impressions of the sun, which comes and goes but does not thereby lead us to think that the interrupted impressions imply a constant annihilation and re-creation of the sun. We disguise the interruptions; we remove them by supposing that the perceptions are connected by 'a real existence of which we are insensible'. This idea of continuous existence which is the product of the imagination is reinforced by the memory of previous interruptions and operations of that propensity to regard them as indicative of continued existence, until it achieves the force and vivacity of a belief. This belief in the notion of an independent and continuous existence has taken such deep root in human imagination, that it is impossible ever to eradicate it 'nor will any strained metaphysical conviction of the dependence of our perceptions be sufficient for that purpose' (*id.*, p.214).

Hume's views on *the causal relationship* are widely known, though the significant place which imagination plays in his analysis is not equally recognised. His argument runs as follows: when we make an inference from cause to effect, from an impression to an idea, we do not do so by examining their essences; it is not a relation of logical implication. The inference is based upon experience, first of all in this way, that we observed in the past a constant conjunction of two species of objects in contiguity or succession, and have experienced that conjunction. The causal inference occurs when either of them appears as an impression or as a memory-image, and the other is supplied by the inference. But the mere repetition of even an infinity of past impressions will not generate a new idea such as that of necessary connection, and so Hume probes further into the nature of the inference, saying, we might think rather paradoxically, that if we can not derive the inference from the necessary connection, we may be able to derive the necessary connection from the inference. If experience produces the idea of necessary connection, it does so either by understanding or imagination. If the former, reason would employ the principle that instances which we have not experienced resemble those which we have experienced, and that the course of nature continues the same uniformly. No one can demonstrate the first of these presuppositions, while the second may be stultified at any time. So it has to be imagination that is the key to the way in which the presentation of one object leads to the thought of the other. A cause is therefore defined as 'an object precedent or contiguous to another, and so united with it in the imagination, that the idea of the one determines the mind to form the idea of the other, and the impression of the one to form a more lively idea of the other' (*id.*, p.172).

As regards *personal identity*, Hume rejects the theory that the self may be the object of immediate consciousness (*id.*, p.252). 'When I enter most intimately into what I call my self, I always stumble upon some particular perception or other, of heat or cold, light or shade, love or hatred, pain or pleasure. I can never catch myself at any time without a perception, and never can observe anything at any time but the perception'. The self on these terms is 'but a bundle or collection of different perceptions which succeed one another with inconceivable rapidity'. Nevertheless Hume does not regard such a conclusion as the end of the matter, though that has been traditionally thought to be where he stands on the philosophy of the self. On the contrary, he acknowledges the existence of what he calls 'a lively belief in the self', and the source of such a belief in the self as an uninterrupted and continuing identity is an activity of the imagination, which moves by an easy transition, through the resemblances, memories, and causal relations of the successive impressions to the bold affirmation that these different related impressions are the same uninterrupted person. This activity of the imagination Hume calls 'fictitious', but he does not use the term pejoratively; rather does Hume seem to be drawing attention to the creative function of imagination, what Kant later was to call its 'productive' nature.

Next, at I.iii.9 (*Treatise*, p.108), Hume extends the application of the concept of imagination to his view of the whole world, and in particular to *historical knowledge*. He mentions the lively quality of the ideas of memory, and adds that we very soon begin to extend them to form a systematic presentation which includes all existing impressions, which we designate 'reality'. But the mind with its propensity for running on (the galley notion again) at the impulse of custom, or by the relation of cause and effect, finds that in so doing it is determined to form a new system which it entitles 'realities'. This principle of what might be called 'universe construction' is the medium which Hume employs in historical construction. 'By means of it I paint the universe in my imagination, and fix my attention upon any part of it I please' (*id.*, p.108). It forms the frame, the *gestalt*, for example, within which I think historically about the city of Rome, its particular government, religion and manners, its foundation, revolutions, successes and misfortunes. Though these are all affirmed to be 'ideas', for indeed what else could they be? — they are nevertheless distinguishable from those ideas which are simply the offspring of phantasy. These thoughts of Hume, appear, on the one hand as regards historical knowledge, to relate to the views of Collingwood to be considered later (*vide infra*, pp.oo-oo), and on the other, as regards the 'universe-constructing' role of imagination to be close to the views of Iris Murdoch on a similar subject (*vide infra*, pp.oo-oo), so that we shall not for the moment pursue the subject farther, but turn to the final connection in which Hume gives us his ideas on imagination.

That is the subject of *sympathy*. Since this concept lies at the heart of much that Hume has to say on ethics, by connecting imagination with sympathy, he introduces imagination to what is for his philosophy, a new circle of relationships, one however with which we are now ourselves familiar from the previous study of the place of imagination in moral conduct (*vide supra*, pp.72ff). Let us then see how imagination works in sympathy. We have all, as Hume has already said, a lively impression of ourselves and of our own person, so that whatever object is related to ourselves becomes endued with that same liveliness. Nature has created us with resemblances (in our generation we call it 'solidarity') with all our fellow-humans, so that to observe in them some passion is to find a parallel in ourselves. But it is the imagination which makes the transition from the passion in the other to experiencing it ourselves. These resemblances extend to manners, character, country and language. By the mere force of the imagination, what begins as an idea may be converted through increase in liveliness into an impression of the passion or the sentiment concerned. Sympathy performs a role analogous in this field to the role played by understanding in our knowledge of external objects. But this act of imagination which is the ground of sympathy is not a passive emotion which suddenly wells up within us, like uncontrollable tears in our eyes. It is creatively interpretative of the state and condition of another person, and with very little extension of Hume's notion, we would arrive at what we were ourselves saying earlier about identification as fulfilling a key role in human moral relations. In the case of our sympathising with the future of a person, there is a fairly complex process at work. First, in order that the process may begin, we have to have some experience of the present circumstances of the person. That in itself is not enough, but it

THE PHILOSOPHICAL DIMENSION   107

is a first condition. Next, we 'must enter into his condition with such a lively conception of it as to make it our own concern; thirdly, this process of entering into the thoughts and feelings of the other person is due we are told 'to a great effort of the imagination' (*id.*, p.386), which is thus seen to be a controllable and the controlling agent in the situation. Fourthly, we are able to entertain pains and pleasures which are neither ours nor even presently existing. So far we have been presenting Hume's use of imagination in his philosophy, without intruding unduly with commentary. But we can not leave the matter there, for there are several aspects of his treatment of our subject which merit further discussion.

To begin with, a general reflection. Hume has suffered over the years from the way on which his expositors as well as his uncharitable critics have sought to categorise him. Those who have known him mainly from his *Dialogues concerning Natural Religion* are swift to call him an atheist, though even that style does not do justice at all to the subtlety of his argumentation, or to the care with which he draws up his conclusions. Others, with a wider knowledge of his work regard him as a sceptic, and they quote his views on the self, or causality, or the nature of the external world and its objects, as these views are expressed at some points in his works, but clearly disregarding the passages which we have considered above. In British philosophy over the past fifty years or so, he has been hailed as the fountainhead of empiricism, at whose waters we must constantly drink if we are to remain fresh and creative in our presentation of the themes of this particular school of philosophical thinking (cf. Roland Hall, *50 Years of Hume Scholarship*, Edinburgh University Press, 1978). Norman Kemp Smith, who was my own mentor in Humean philosophy as in so much else, used to find in the influence of Hutcheson upon Hume and in Hume's views on 'natural belief' the corrective to some of the more brash approaches to this Scottish philosopher. It was not however until I personally became interested in the whole theme of this study that I became aware of the pervasive part which imagination plays within Hume's thought, as we have tried to demonstrate in the elaboration of its presence in the above five areas. In fact, to play down or to neglect the place of imagination in the Humean philosophy is to that extent to begin to misrepresent it. From our point of view, however, the value of Hume's contribution is his demonstration of what I earlier called the bread-and-butter jobs that it can perform; and Hume demonstrates just how wide-ranging these are. He takes the concept out of the world of phantasy, and, though we may feel that he does not sufficiently consider its aesthetic role, he gives it an honourable status within those fields of thought within which traditionally philosophers since Plato and Aristotle have operated.

What Hume presents is an account of the central place which imagination plays in the cognitive, and we should add, the moral, relations in which we stand to the world and to society around us. He illustrates his case by examining its applicability to the central themes which any philosopher in the western tradition would regard as important — the nature of the self and our awareness of the self, the character of the causal relationship, the status of the external world and its continuity and objectivity, and the very nature of knowing itself. In selecting his list Hume has not gone for easy options. In this connection it is worth saying that what comes out of his treatment of imagination as applied to these different topics is a singularly coherent presentation, so uniform, in fact, as to constitute a major principle in Humean philosophy.

I have made reference to the fact that Hume sees imagination as being present not just in our cognitive relations to the world but also in our moral obligations to other members of society. His treatment of this part of his philosophy is relevant evidence in the matter to which reference was recently made, namely, the categorisation of Hume's philosophical stance. It is customary to class him, as regards his ethical theory, with the 'sentimentalists', those who believed that moral discernment and decision was a function of the so-called 'moral sense'. Clearly, from some of his statements about sympathy, he has affinities with that group. Yet, when we look more closely at his exposition of the subject, we see that what he says is remarkably akin to 'identification' as we described in our account of the ethical dimension of imagination. Not only does he provide us with an anatomy of identification, but he also sets out a map of how we ought to proceed if we wish to employ it as an ethical guide for our own thinking and action.

Finally, the question might be asked why it is that, if imagination fulfils the role in the philosophy of Hume for which I have been arguing, it has not been given such prominence in the past. A possible answer would be as follows. Interpreters of Hume as of the empiricist school of philosophy in general, have difficulty in distinguishing between epistemology and psychology in his writing; and their difficulty is understandable. He writes of many of the subjects that concern psychologists — ideas, images, memory, self-awareness and so on. The difference is that he is not so much interested in the furniture of the mind as in the validity of its activities, the truth of its judgments, the character of its relations to reality, to past and the future, as well as with the analysis of logical steps by which it moves to its conclusions. The consequence has been that the contribution which Hume's analysis of the concept of imagination might have to make to the epistemological discussion has gone by default, having been relegated to another discipline. This consequence does not surprise me. Something similar happened to Kant's account of imagination and the part, the all-important part, that it has to play, in his opinion, in our awareness of the world of objects. In Kant's case there has been no compensating appearance of his ideas on imagination in psychology; in his case, the compensation has appeared in the field of aesthetics. (Note should be taken at this point that Kant will be omitted from this philosophical review of the concept of imagination, for the reason that an essay of mine, 'Theology and Human Imagination: New Help from Kant', appears in a collection of essays entitled *Religious Imagination*, edited by James P. Mackey, Edinburgh University Press, 1986).

Collingwood

Collingwood, in his *The Idea of History*, Oxford, 1946, introduces us to a rather different application of the concept of imagination, namely, to history and historiography, an application which is of tremendous importance to us in our endeavour to see how this concept may assist us in understanding the Christian faith and Christian theology. We observed, when discussing George MacDonald (*vide supra*, pp.15f), that for him one of the most important features of the imagination is the 'intellectuo-constructive' role which it plays in the process of historiography; for it interprets the significance of movements within history, analyses the motives which activate the agents of history, elaborates shreds of evidence into completed patterns, and even discerns the will of God among the volume of happenings and actions in the material before the historian. There are remarkable similarities here which anticipate the much more sophisticated theories of Collingwood, as we shall see. It is interesting, too, that Bultmann, who never showed any great interest in British philosophy compared with his absorption with continental existentialism, should have paid a special tribute to Collingwood in his Gifford Lectures, *Eschatology and History*. Collingwood leads us into his account of the place of imagination in history and historiography, with a description of what he regards as three erroneous methods of writing history. First, there is the 'scissors and paste' method, where the historian has before him, or, by his industry discovers, certain 'authorities', that have written about the subject of his research. These authorities he accepts as veracious, and incorporates their statements, usually suitably paraphrased and to that extent somewhat disguised, into his account. This might be called the commonsense view of historiography; but that it is not an accurate description of what goes on in this field, can be seen from the following circumstances: first, that the historian makes his own selection of authorities, and of the statements that they have supplied, which selection introduces criteria of historical relevance and therefore of truth, which are not contained within the authorities themselves; next, that the historian interpolates material which is not to be found in the authorities at all; that is, he is constructive and not merely mimetic; and finally, he is critical of his authorities, on occasion convicting them of ignorance or misrepresentation, so that he is very far from treating them, as one might have been tempted to think when he was calling them 'authorities', as literally inspired. The second erroneous method of historiography which Collingwood castigates, he names as 'critical history', which is a refinement of the 'scissors and paste'

THE PHILOSOPHICAL DIMENSION 111

approach. Historians practising this method substitute for the authorities of the last example the idea of 'sources', whose reports and evidence are not accepted at their face-value, but questioned and carefully scrutinised. Once such questioning and scrutiny have been completed, however, and the credentials of the sources have been established; then the sources become effectively authorities once again, and the techniques of the previous method are adopted. The critical method then becomes vulnerable to the objections of the first method, insofar as it does not elicit the constructive features of the historian's method, a central theme, as we shall see, in Collingwood's own analysis. A further, third, unacceptable form of historiography is designated rather strangely 'pigeonholing' which is an activity indulged by frustrated historians, who refuse to agree that the previous two methods exhaust the possibilities. 'Conscious of having brains, they feel a laudable desire to use them' (*op. cit.*, p.264), and they extrapolate supra-historical structures in which to systematise their findings about history. The investigator discovers what Reinhold Niebuhr used to call 'tangential meanings', partial patterns which he then constructively completes. Examples of such historiography would be St Augustine in his *de Civitate*, O. Spengler, *The Decline of the West*, or A. Toynbee, *A Study of History*.

In distinction from these three fallacious conceptions of historiography, Collingwood formulates the notion of '*a priori* imagination' as the key to a new attitude to authorities. From the easily established facts that the historian does not simply join together the statements made by the authorities, but actually endeavours to relate them meaningfully to one another; and that he often finds himself making statements that are nowhere to be found in any authority; Collingwood concludes that one of the most characteristically historiographical activities is that of *interpolation*. The interpolative process has two components: it is *a priori*, and it is imaginative. In saying that the interpolation is imaginative, Collingwood is not saying that it is fancifully inventive of facts which lie beyond the justification of the evidence, or the sound judgment of the historian or the other experts in the field. The interpolation is constructive, weaving a web of valid conjecture, so that the part of history being studied becomes meaningful and intelligible to the historian and his colleagues, and exhibits a pattern of continuity. On the other hand, when he says that the interpolation is *a priori*, he means that it is necessitated by the evidence which is relevant to it; such evidence leaves room for no alternative possibility. He goes even farther, adding that this process of interpolation takes place within what he calls 'the idea of history', an innate idea which is an essential constituent of the human psyche. What had previously been known as authorities but are now reduced in status, or what would still be called evidence, will appear as 'nodal points', which are not to be accepted as final and authoritatively prescriptive, but as starting-points concerning which the historian asks such questions as, 'What do they mean?', 'What are the implications of this evidence?', 'What attitudes does it reveal in the mind of the speaker, or the agent, what motives, what stratagems, what plans, what deceits?'

So far Collingwood has been speaking about historiographical method. He turns next to the question of how we come to know history, and *a priori* imagination is of the essence of the process. The historian re-enacts in his own mind the inner, the thought side of past events in which purposive beings have acted. It is therefore necessary to re-enact the past to know it. At this point, the plot thickens somewhat, as Collingwood states that all thought as an act has both a subjective and an objective aspect. As subjective it occurs within the experience of a given individual, and thoughts never exist other than in such a context. On its objective side, an act of thought can be an object of thought, and it becomes such, not by being set ready-made before someone, but by being re-thought, that is, re-enacted. The presence of thought in a flow of consciousness, Collingwood calls its immediacy. But thought can transcend its immediacy in one context and revive itself in a different one by being re-enacted there. It is *a priori* imagination which is at the heart of this process.

It would be out of order for me in the present context to endeavour to assess the stature of Collingwood as an analyst of historiography or as an interpreter of philosophy of history. Suffice it to say that that stature is implicitly recognised in the important place assigned to the continuing examination of his theories in *Form and Substance in History* (edited by L. Pompa and W. H. Dray, Edinburgh University Press, 1981), which is a critique of writing in the analytic philosophy of history since Collingwood, mainly with special reference to the work of W. H. Walsh, in whose honour the volume has been written. But from the point-of-view of our particular interest in the place of imagination in faith and theology, we may make three comments on Collingwood's theories. First, his interpretation of the nature of historical method comes with a certain freshness, not to say novelty, in the wake of the excess of what can only be called historical scepticism which has dominated both protestant theology and biblical criticism for many decades now, in some cases for a century. On the theological side, it had its chief and most influential source in Kierkegaard, with his categorical assertion that 'an approximation is the only certainty attainable for historical knowledge' (*Unscientific Postscript*, E. T., Oxford, 1941, p.31). This book though written and published in Copenhagen in 1846 did not really begin to influence European theology until the 1920s and British and American theology until the 1940s with the extensive translation of all of Kierkegaard's works into English. Such historical scepticism is shared by writers who in so many other ways differ greatly from one another; I am thinking of Barth and Brunner, Tillich and Bultmann. But such scepticism was not confined to theology, for it appeared in the biblical field, particularly in New Testament studies, with increasing uncertainty being created as to how much was to be truly known of what had once been called the 'historical Jesus'. Here, however, the sceptical method was given a rich variety of expressions through a whole range of critical methods, named after the dominant feature of the method, such as, source-criticism, form-criticism, redactor-criticism, audience-criticism, and so on. Incidentally, what was largely lacking from that situation was a thorough-going analysis of the historical critical method implicit in more or less all of these forms of criticism. Collingwood offers us an alternative to the pervasive scepticism to which I have referred; and it comes not from casual reflection upon what historians do, but on a careful analysis of how he himself has written history. The historical method which Collingwood advocates endeavours to understand the past and to describe it, by a process of re-enacting the intentions and motives of the agents, and constructing an image of their character as it expresses itself in their actions. Such an approach to the study of history has more to give to the understanding of the biblical scriptures than had the older sceptical criticism, seeking to view the subjects of the scriptural narratives as persons, rather than as collections of texts which have to be validated before they can be further employed. What Collingwood has said about method may, secondly, be extended to throw light upon the nature of Christian faith. Such faith has a great deal to do with the appropriation of the biblical accounts of events in the history of Israel, the life of Jesus Christ and the mission of the Christian Church. As Kierkegaard so often warned, simply to be aware of these events as past history is not to be a believing disciple. We have to enter into them with that closeness of relationship which he termed being 'a contemporary disciple'. That contemporaneity, if we follow Collingwood, will be achieved by an imaginative re-enactment of the events of which we read in the biblical records. In other words, what Hume had regarded as essential to sympathy in our relationship to someone in distress, what we ourselves were holding is the generating characteristic of identification with another person, in joy or in sorrow, is that same imaginative process of self-projection into a situation beyond us and perhaps in many ways different from anything that we have ever known, which Collingwood is here describing. The difference is that Collingwood is relating this process of identification to the *past*; and that is where his reference is so helpful for our appreciation of how faith is in human terms made possible and effective.

Finally, since Collingwood employs the concept of imagination for his interpretation of historical method, it has been suggested that his subsequent presentation must at times therefore border on phantasy. An immediate reply to such a caricature is to draw attention to the fact that Collingwood rarely employs the term imagination by itself, preferring mostly to prefix the qualifying idea of *a priori*. This idea is the subject of much analysis by Collingwood scholars; and there is no doubt that much of the difficulty arises from the acknowledged fact that he uses the term in several senses. For example, it is a psychological *a priori*, a capacity of the human mind, this ability to know the past imaginatively. It is a temporal *a priori*, in the sense that ante-dates the occurrence of any actual knowledge of the past. It is *a priori* in the sense in which Kant used that term of the categories; it is the condition of the possibility of our knowing the past. But there is another sense which applies, and it is the one which concerns me now; it is a logical *a priori*. The construction which imagination effects has to conform to the logical requirements of the field of study to which it is directed. It has to be recognised as determined by such evidence as is available. In this connection, Collingwood used to place considerable reliance upon the judgments of colleagues working in the same field. The construction has to have the quality of necessity. I find this view of his particularly salutary in our own field of theology. We, too, could be charged with indulging in phantasies in introducing imagination into theological construction; but again following Collingwood, we reply that such a method does not expect to be exempt in any way, when it brings forward its substantial statements and its conclusions, from the rigours of the logic to which all theological formulations must be exposed. Theological positions reached through the employment of the imaginative process must conform to the criteria which all theology has adopted.

Sartre

The influence of Existentialism upon modern theology has usually been traced to either Kierkegaard or Heidegger, but not nearly so often has Sartre been regarded as a force of equal intensity. Nor, for that matter, has Sartre himself been taken to be particularly sympathetic to Christianity. Yet obliquely, his understanding of imagination and images, as set forth in *The Psychology of Imagination*, E. T., 1972, is a very valuable contribution to our awareness of the nature of the Christian faith. I doubt if anyone has done more than Sartre in modern times to re-habilitate the concept of imagination, or to open up a place for it in the presentation of the faith. But first we must examine what Sartre says on the subject of imagination, appreciating its importance for him also when he claims, 'We restore to imagination as such, which disappeared when psychologists ceased to believe in faculties, an importance which cannot be exaggerated, as one of the four or five great mental functions'. We shall begin by summarising what he calls the four characteristics of images. The first characteristic of the image is *consciousness*. He rejects as error two ideas: first, that the image is *in* consciousness; and secondly, that the object of the image is *in* the image. He describes as the 'illusion of immanence' the supposition that our images have the characteristics of their objects, which leads to our thinking that our consciousness is populated with small likenesses which are images. Having perceived a chair and closing my eyes, I produce an image of the chair; but the chair now occurring as an image can no more enter my consciousness than it could do when I was perceiving it. The word 'image' on the contrary connotes the relation of consciousness to an object, the way in which consciousness presents the object to itself. It is better to speak of 'consciousness of Peter as an image', or 'the imaginative consciousness of Peter' than of 'the mental image of Peter'. The second characteristic of imagination is the phenomenon of *quasi-observation*. There are three types of consciousness in which an object may be given to us: perception, conception and imagination. In perception, the object enters my consciousness in its completeness, yet I see it always and only in perspective; it appears as a series of profiles and projections, and it has an infinite number of relationships to other things as well as between the elements of the thing. These relationships constitute the essence of the thing. In conception, I can think of the six faces of the cube and its eight angles at once. In imagination, the image appears in profiles but the image presents itself as complete in appearance. 'The image is a synthetic act which unites a concrete non-imaged knowledge to elements which are actually more representative. The object presents itself with a certain opacity and as requiring a number of synthetic acts for its apprehension. So it gives the impression of being an object of observation, that is, of quasi-observation'. Sartre is here endeavouring to expand on his previous insistence on the difference between images and their objects, having said that images are not likenesses of them in the mind. He seems to recognise the source of the error by acknowledging that images are profiles and projections, as are perceptions; but while the object of perception remains 'always and only' in perspective, the image is a synthesis of non-imaged knowledge and representative elements, which Sartre seems to think can not be called a 'likeness'. The third characteristic of imaginative consciousness is that it posits its object as *nothingness*. Every act of consciousness posits its object in its own specific way. Perception posits its object as actually existing, so that, when by contrast Sartre says that the imagination posits its object as 'nothingness', he means that imagination posits its object as non-existent in the world of sense-perception here-and-now, but nevertheless as existing in other ways, as absent in some other place, or indeed in our own thought. Imagination, that is, can suspend belief and not posit the object as existing at all. The fourth characteristic of imagination expresses a similar point, namely, spontaneity. Echoing a widely accepted, if not altogether accurate epistemological theory, that perceptual consciousness seems to be largely passive, the *tabula rasa* registering stimuli, the imaginative consciousness is creative, though its creativity does not extend to positing its object as actually existing. The third characteristic would firmly over-rule such an eventuality.

There is a refinement to what Sartre has been saying about the image which brings him closer to the traditional understanding of the subject, as expressed by Hannay and Dame Mary Warnock (*vide supra*, pp.90f.). On a number of occasions, but never in sustained expository passages, Sartre refers to an entity called an 'analogue' (for example, *op. cit.*, pp.26, 31, 64, 102, 105f., 217) which functions as the equivalent of the perception, or the representative of the absent object which is being envisioned. At p. 102, he refers in these terms to our imaginary consciousness of the Pantheon, remarking that the imagined monument though absent, is nevertheless present, through the good offices of the analogue, which, while representing the sensible qualities of the Pantheon, does not possess them. Through the transfer to the analogue of the qualities of the thing it represents, 'a miniature Pantheon is constructed for the imaginary consciousness' (*id.*, p.102). This circumstance Sartre takes to be the source of what he had identified as the 'illusion of immanence' in defining the first characteristic of images. As so described, the status of the analogue is rendered somewhat ambiguous; for Sartre may mean one or other of two things, namely, either that the analogue only seems to be functioning in the mind as the representative of the object envisioned, but is not really there at all, and the person imagining it is to that extent deceived; or alternatively, the illusion for all its illusoriness, and appearance of being a miniature of the Pantheon, nevertheless exists as such for the imaginative consciousness. Clearly, there are unresolved problems here for the interpreters of Sartre, and we can not delay further over them, beyond acknowledging the sensitivity of Sartre to the subtleties of the imaginative process which we have perhaps been tending to oversimplify.

Something of that same subtlety, harsh critics might call it equivocation, appears in a further pair of contrasted views. On the one hand, he affirms that 'the act of imagination is a magical one. It is an incantation designed to produce the object of one's thought, the thing one desires in a manner in which one can take possession if it. The object obeys the order of my consciousness' (p.141). This manipulable character of the media of imaginative consciousness has, of course, been the prime source of the total distrust in some theological circles of my acknowledgement of a place for imagination in theology, though I would be inclined to read what Sartre has here written, as being a rather extravagant way of stating what he has now said frequently, namely, that there is in the imagination a strong element of spontaneity, what in fact George MacDonald was calling its creativity. In any case, we have to balance that view against the second emphasis he makes concerning the relationship of imagination to the world. In the 'Conclusion' (*op. cit.*, pp.201ff) Sartre, acknowledging the Kantian nature of his interest and eventual formulation, affirms that imagination is a condition of the possibility of ordinary consciousness of the world. 'Imagination, far from appearing as an *actual* (which in the original French would have meant accidental, fortuitous and occasional) characteristic of consciousness turns out to be an essential and transcendental condition of consciousness. It is absurd to conceive

THE PHILOSOPHICAL DIMENSION

of a consciousness which would not imagine as it would be to conceive of a consciousness which could not realise the cogito', that is, the act of thinking which Descartes regarded as the demonstration of one's existence (*id.*, p.210). Another way in which Sartre states this point is to say that the imagination posits the world in its totality, and as a synthetic totality.

Briefly, we may consider three directions in which Sartre develops this theory about the relation of imagination to the world. First, so far is Sartre from antithesising the imaginary and the real, that he presents them as indissolubly integrated with one another, saying on occasion that the imaginary is the transcendental condition of the world, and on another occasion, that the imaginary 'appears upon the foundation of the world' (*id.*, p.211). Here what Sartre is saying is highly reminiscent of Kant in the *First Critique*. Secondly, Sartre extends the role of the imagination in relation to the world, by saying that the imaginary represents at every moment 'the implicit meaning of the real', and constantly endeavours to make that meaning explicit. Here we have the affirmation of the continuity of the imaginary and the real, with the real world implying 'a hidden surpassing towards the imaginary'. Thirdly, Sartre is now in a position to propound his theory of the nature of freedom. If it is the role of the imaginative consciousness to posit the world as a synthetic whole, that same consciousness by taking the step of imagining can escape from that world, and this going-beyond-the-world is the very essence of freedom. It would be impossible for a consciousness that was not free. 'An image enmired in the world is that of psychological determinism'. Therefore, Sartre argues, the very ability to posit the world, and to be the condition of its constitution, implies the ability to go beyond it and thereby to posit one's own freedom of that very world. In a word, imagination is the form which human freedom takes, and it expresses itself in both the constitution and the negation of the world. Sartre concludes that 'imagination is not an empirical and superadded power of consciousness, it is the whole of consciousness as it realises itself in freedom' (*id*, p.209).

Let me conclude this review of Sartre's views on imagination with the following comments: First, it is only proper at the start to acknowledge the illumination which Sartre casts upon the subject, both by the range of his analysis and by the clinical method which he employs. Though the title of the book proclaims it to be a work in the field of psychology, it would be more correctly classified as phenomenology, thus covering both epistemology and philosophy of mind. The movement from one discipline to the other is perhaps the main source of some of the more obscure passages, and descriptions. Though Sartre's presentation was absolutely without intent to assist the theological enquiry, nevertheless there is much here that is bound to be helpful. For example, secondly, while there is a certain unclarity about the role of the analogue in Sartre's epistemology, and especially about the relation of the image to the analogue, he is fairly definite in his description of the role of the image in relation to the object which imagination is envisioning. They mediate the object to consciousness. When he similarly describes the role of imagination itself, he affirms that it is a way, in addition to, and quite distinct from, perceiving and conceiving, of consciously positing an object. Put more simply, it is one way of thinking about an object, distinguishable from the other two, and as much a part of our everyday thinking as they are. It is the mediating consciousness by which the absent is envisioned as present, or the non-perceived is thought about and talked about. When we so describe the function of imagination we are close to the role which previously we ascribed to the parabolic thinking which the New Testament reported Jesus as using so extensively in order to present his message. If we wish to think about the Kingdom of God, say, we can not do so by perceiving it, for it is not that kind of entity. Nor do we arrive at the truth about it, if we endeavour to conceive it, though theology has endeavoured to do so on many occasions by developing an eschatology. The method which Jesus demonstrated for our direction was that of the wide-ranging series of parables; and the way in which we entertain these parables is exactly by employing that imaginative consciousness which Sartre has so carefully outlined for us. Incidentally, it is noteworthy that Sartre accepts the same threefold division of the ways of knowing which Hume had adopted, namely, of perceiving, conceiving and imagining. The remarkable fact, to which we shall be devoting more space in the next chapter, is that

THE PHILOSOPHICAL DIMENSION

religious epistemology has consistently ignored the third of these, despite the strong evidence, as has just been indicated, of the New Testament which suggests quite a different reading of the situation. Thirdly, Sartre's association of freedom with imagination may owe something to the circumstances which obtained in France during the years of World War II, when the outward manifestations of freedom, of public and unrestricted expressions of political opinion, of untrammelled movement wherever one chose to go, of association with groups of likeminded citizens, and so on, were brutally forbidden. Yet Sartre's point, however time- and circumstance-conditioned, remains valid in days of greater political liberty. For freedom has in the end to be an inward condition of the spirit or the mind; otherwise, even in gaining the whole world we shall lose our own soul. In fact, it is how we shape and discipline that inner world that determines how we avail ourselves of these other, more external freedoms, and saves us from abusing them, or more tragically, of turning them into new forms of bondage. Fourthly, I gain great assurance from Sartre's insistence that imagination is continuous with the real world, that it has a very special part to play in the construction and maintenance of that real world, and further that imagination seeks to make explicit the implicit meaning of the real. Sartre can not be criticised for failing to implement such claims, and his resolution in so doing strengthens the case for the sort of probing we have been seeking to do in relation to the place of imagination in theology. Indeed, if the human mind works generally in the way in which Sartre portrays it here, it will be very surprising if it ceases to operate in a somewhat similar way in Christian faith and theology. Also, if we are speaking about making explicit the meaning of the world about us and the society in which we live, then we are referring to one of the main functions of any religion and certainly to a major accomplishment of the Christian faith; and by implication to the place which imagination might have in such a process. Of course, another way of describing what imagination does, in making explicit what is implicit in the meaning of the real is to ascribe to imagination an interpretative role in our relation to reality. At that point, we are close to yet another of our later interests, the place of imagination in the epistemological situations in which we find ourselves.

Warnock and Murdoch

Dame Mary Warnock towards the end of her book (*Imagination*, pp.196ff.) gathers up in one comprehensive statement the view of imagination which has evolved as she reviewed the writings of those who have dealt with the subject in modern times. There is, she holds, a power in the mind, which operates in our everyday perception of the world around us, in our thinking about objects and persons when they are absent from us, so that we endow them with a kind of presence; which enables us to perceive significance in the world around us, as we have just been observing Sartre to say, to interpret it and to communicate that interpretation to others, 'for them to share or reject'; and which finds expression in, and enables appreciation of, the work of the creative artist. This compendious statement is so interesting that it will repay further reflection, particularly at the following points. First, when she speaks of 'a power in the mind', she must not be understood to be reverting to some kind of medieval faculty psychology, and so to an extra sense which some people may possess, while others are less unfortunate. Two comments may be made in explanation. The one is that since Warnock does not go on to list other 'powers of the mind' each operating independently of the others, we need have no fear that the medieval psychology is being revived. In fact, no such list appears, and so she is exempt from such a criticism. The other comment is that clearly when she speaks of 'a power in the human mind', she means 'a power *of* the human mind' and is referring to the mind *as a whole*, acting in a certain way, thinking of certain things or persons in their absence, or of certain aspects of them, or of some objects as carrying a significance beyond that which can be 'read off' by simple sense-perception. Secondly, we observe that this single term is employed to comprehend within its operation, a whole variety of mental activities, and it is of the essence of imagination to express itself in a wide range of psychological phenomena. It is my conviction that no apology requires to be made for employing one term to refer to that wide range and variety. We do the same when we apply the term 'reason' to an equally vast range of mental activities, as we do the term 'perception' to the multiple operations of the senses. Now, just as we acknowledge the range of meanings of 'reason' and refrain from caricaturing it by identifying it with one of its less acceptable forms, for example, rationalisation; so, in fairness, we ought to reject attempts to equate imagination with phantasy, and give it the credit of its valid forms. Thirdly, it is noteworthy that Warnock comes down firmly in support of recognising what we called 'the bread-and-butter' jobs that imagination does within the ordinary activities of perceiving what goes on around us, being in relation to our neighbours and communicating with them, and evaluating moral qualities and making decisions. In fact, it is most important to stress once again that imagination is no optional intellectual extra; it is of the very stuff of human knowing, feeling and willing. As such, it has been given a place in both epistemological and moral theory by Plato, Hume, Kant and Sartre, to mention but a few from a continuous tradition within philosophy. Fourthly, the next element in Warnock's delineation of the nature of imagination is of enormous importance to us in our investigation of the theological potential of imagination, namely, her view that it enables us when thinking about objects or persons when they are absent from us, to endow them with a kind of presence. We met this notion of presence-in-absence, as was noted, in Sartre, who used it extensively in his analysis of what happens when we look at a photo of an absent friend; it has featured frequently in the theology of Professor John Zizioulas; and Warnock is saying that this kind of thinking is a function of imagination. The religious epistemological value of the idea is that it provides a way of conceiving how we come to know God, who is not visibly present and is to that extent 'absent'; who can not be reached through the processes of ratiocination; who for many saints across the ages and own their own testimony has withdrawn from them for a season so that they seem to be living in 'a cloud of unknowing'; but who nevertheless still exists and is real. Imagination can conceive of that God as present-in-absence. In fact, imagination may well be the form which faith takes in such circumstances. Fifthly, it is through imagination, too, that we are able to interpret what goes on in the world around us. Wittgenstein introduced into philosophy a notion which has fairly caught on in theology, the notion of 'seeing-as', and Warnock is saying that the mind's ability to see the world *as* the creation of a loving God or, for that matter, *as* the meeting-place of irrational forces, is the work of the imagination. Finally, Warnock assigns to imagination a significant role in the communication process, in which the artist or the poet or the novelist seeks to make available for others, experiences which are private to himself or herself. By communicating their insights to the world at large, by thus publishing and in a sense universalising them, they invite validation or rejection of their experiences or ideas or presentations in picture, drama, or verse. Art is to be shared, and not solipsistically to be enjoyed by its creator; and the break with solipsism comes in the act of communication, the point where imagination becomes all-important in the choice of the media and the appropriate images.

Turning now to Iris Murdoch, we dwell on a valuable contribution which she has made to our theme, almost, it might seem, in the by-going, for it appears in a review of Stuart Hampshire's *Freedom of the Individual* (*Encounter*, xxvii, July 1966). There she describes imagination as 'a type of reflection upon people, events, etc., which builds detail, adds colour, conjures up possibilities, which go beyond what could be said to be strictly factual'. Here we have something of Hume's 'galley' concept, the idea of the mind running on beyond the immediate deliverances of the senses, and beyond positions justified by reason, to conclusions which it invests with factual status, and which in certain cases come to govern our behaviour. The suggestion that imagination supplies detail and colour to life is a particularly attractive contribution to our dossier on the subject. Murdoch supplies yet further fresh direction to our thought, when she points us to the social and inter-personal importance of imagination. In our relations with others, we are influenced by beliefs about them which are imaginative constructions, and are sometimes consciously determined by our wills. The emphasis upon wills is significant, for we have not so far come across this notion that

THE PHILOSOPHICAL DIMENSION

imagination has sometimes to be stimulated by a very definite decision on our part that it must operate in a certain direction. Part of her earlier objection to the way in which Hampshire handled imagination was that he relegated it to the passive side of the mind. So, she continues, 'we have to attend to people; we have to have faith in them, and here justice and realism may demand the inhibition of certain pictures, the promotion of others. In fact, through the imagination, we all of us construct our-world; we choose to live within this world, partly private, partly fabricated'. This world contains the value-elements of our lives, the controls that affect our decisions, intentions, choices, as well as our friendships, and our long-term plans for our lives. It is the imagination which, as it were, allows and enforces the colour and magnetism of these values to control our decisions. 'The world *we* see already contains our values, and we may not be aware of the slow, delicate processes of imagination and will which have put those values there'. The possibilities of a theological transcription of this role of imagination become attractive, once we realise that what Murdoch says of imagination as constructing an envalued world could be said with even greater force of the way in which, under God's grace, that same imagination may build up a system of Christian convictions and a life interrelated with others in bonds of love and goodwill and sincere appreciation. The will to which Murdoch ascribes such importance in the construction of our-world is, in its Christian form, the decision made by faith in response to God's approach to us in Jesus Christ.

So positively helpful have Warnock and Murdoch been that it may seem somewhat ungracious to raise a question, but what I did miss from both of them was a point made by Plato when he dealt with the 'mathematicals'. He held that imagination operates at two levels of the Divided Line — at the first stage, as constituting the only possible knowledge for the dwellers in the cave. It re-appears at the third stage, the lower form of *episteme*, namely, *dianoia*, understanding which works with the 'mathematicals'. Understanding thinks about the true Forms, but does so, not directly as *noesis*, the very highest form of knowledge does, but by means of images or symbols, very much what modern philosophy would call 'models'. Plato reminds us that, in addition to the activity of imagination in ordinary perception, in the aesthetic appreciation and creation of the poets and dramatists, or even as Murdoch herself suggested, in the construction of our-world, with its values, concerns and friendships; the imagination works in the field of scientific enquiry and the construction of scientific hypotheses, as George MacDonald claimed and as Max Black and Ian Barbour also held. Einstein's one-field theory is an act of creation of the highest kind, and is no less so for being conditioned by and made possible on the assumption of, a great deal of firmly based and carefully and rationally worked scientific calculations. In such an account of the matter, we are also emphasising the organic unity which exists between imagination and reason within the wholeness of the human psyche.

CHAPTER 6

IMAGINATION
ITS PLACE IN METHODOLOGY
AND EPISTEMOLOGY

Methodology

It would be surprising if, having explored the role of imagination in so many areas of theological subject-matter, we did not encounter it also in theological methodology and epistemology. For method and matter in this, as in most disciplines are deeply involved with one another. As we move into what is a part of the theology more rightly described as philosophy of religion, we shall be aided by the discussion just concluded of the way in which some of the philosophers have dealt with imagination. In our new departure, the form of imaginative activity upon which we shall be concentrating will be the creation and employment of *models*. This notion has been widely employed by theologians in the past two decades, notably by the late Ian Ramsey in *Models and Mystery*, Oxford, 1964, and *passim* throughout his writings. For my own part, I found it convenient to compare the different Christological theories with one another as 'models' when writing *The Shape of Christology*, SCM Press, London, 1966. But the term has a primary scientific reference, and consequently has been given considerable attention in philosophy of science, so that any subsequent application of it to theological subject-matter is to a degree analogical and derivative. It could even be said that in theological methodology we employ models as 'models'. It may therefore help our own theological discussion if we pause a little to consider the primary significance of the term, following the guidance of two popular exponents of the subject.

The first, Max Black (*Models and Metaphors*, Cornell U. P., 1962, pp.219-243) enumerates five types of models, commenting upon the nature and status of each. First, developing from literal cases of miniature models such as the ship at one time displayed in the show-case of a travel agency and labelled 'Model of the Queen Mary', there is the *scale model*, which includes all likenesses of material objects, systems or processes, which preserve the relative proportions obtaining in the originals. The scale model is a representation of the real or the imaginary thing for which it stands; it exists so that the user can 'read-off' properties of the original from it, and the activities of the model can be regarded as the basis for the prediction of how the original will be expected to behave. 'We try to bring the remote and the unknown to our level of middle-sized existence' (*op. cit.*, p.221). Secondly, when models are constructed which involve a change of medium, the example given by Black being a hydraulic model of an economic system, we have *analogue models*. Their purpose is to reproduce as faithfully as possible the structure of 'web of relationships' obtaining in the original. Scale models aim at achieving identity between model and original; whereas analogue models are based on a point-to-point correspondence between the relations which structure the model and those in the original. 'Isomorphism' is the dominating principle of the analogue model; it is ikonic of its original, but in a much more abstract way than the scale model. Both are required to conform to their appropriate rules for translating the terminology applicable to the model into true language concerning the original. It may be observed that Black's distinction between scale model and analogue model runs very close to Sartre's distinction between analogue model and image, though the analogue plays a totally different role in the two sets of correlates. Thirdly, *mathematical models*, popular in the social sciences, consist of the projection of the original field and the relations found within it on to the abstract domain of sets and functions found in mathematical theory. In the original field certain variables are identified; hypotheses are formed concerning putative relationships between these variables; considerable simplifications are introduced for the sake of easier formulation and manipulation of the variables; a so-

METHODOLOGY AND EPISTEMOLOGY

lution in the form of a mathematical equation is sought; and finally, extrapolation in the form of examinable and verifiable consequences in the original field is executed. The result expected from the use of such models is a form of explanation that demonstrates which kinds of function fit the data; but this result can never be stated in the form of a causal explanation. Fourthly, *theoretical models*, Black says, are treated by some scientists as expository or heuristic devices or fictions, 'as if' thinking; and by another group of scientists as describing the scientific entity as it is. Using language appropriate to the model, the latter scientists work not by means of analogical argument but through and by means of an assumed underlying ontological analogy between the model and its original. When, however, the model is regarded as a heuristic fiction, it is usually drawn from a relatively unproblematic domain, consisting of well-known objects, materials, mechanical systems and structures, and employed to explain certain given but problematic data, or to extend knowledge or conjecture. The fifth class of models are not so much models though they function like them, and are known as *root metaphors* (*op. cit.*, pp.239ff). A person selects one area of his life and tries to comprehend the rest of his life by reference to it. He constructs a list of categories, a basic analogy or root metaphor, and uses these concepts to interpret the whole of his life. Some root metaphors are more fertile, and have greater powers of expansion than others. These could be called 'conceptual archetypes', which include key-words and expressions with statements of their inter-connections and paradigmatic meanings, in their fields of origin.

From our point of view, the all-important statement made finally in Black's exposition is that this whole process of model-making and archetypal construction calls for the exercise of the imagination. The selection and construction of the model, in whichever of its forms, *precedes* the processes of detailed rational deductions and precise mathematical verifications, and is not derived from them *a posteriori*. The selection of the model is an act of perceptive imagination, based upon a very thorough knowledge of the domain in which the problem to be solved, or the predictions to be made, actually exist, but not in the first instance derived from it, by any means of sustained logical argumentation.

The second author in the field of philosophy of science whom we shall consult is Ian G. Barbour (*Myths, Models and Paradigms*, SCM Press, London, 1974), who presents a classification of models somewhat similar to that of Max Black, though different enough to throw further light on our subject. Here is his list: First, *experimental models* include scale models, working models, and analogue models, in which the features of one system are so constructed as to simulate the behaviour of another system, projected into another medium. Secondly, *logical models* are sets of entities which meet the formal requirements of the axioms and postulates of a deductive system. Thus, Euclid employed a set of points and lines as the logical model for the enunciation of his axioms and theorems. Here we have Barbour endeavouring to describe the situation to which Plato was referring in his theory of the *mathematika*. Such models neither imply nor are expected to imply any reference to real entities; they form *a priori* conceptual systems. Thirdly, *mathematical models* are positioned between the first two, being 'symbolic representations of quantitative variables in physical or social systems' (*op. cit.*, p.30). The relationship between the model and its original, for example, between a mathematical equation and supply and demand in economics, or between a computer model and transport flow, or military logistics, lies solely in the similarity of formal structure which obtains between the two, and does not extend to material or physical resemblance of any kind. The most common purpose to which mathematical models are put is that of predicting how the original system will operate, under a variety of conditions. Fourthly, *theoretical models* are much better explained by Barbour than by Black. 'These are imaginative constructs invented to account for observed phenomena' (*ibid.*) where an analogy is drawn from familiar circumstances or processes, and used not so much for prediction of operations within the original system, as for the provision of a theory which yields an interpretative description of it. Additionally, the theoretical model thus based upon analogy enables correlations to be set up between elements in the original system. Of this fourth class of models Barbour affirms that their 'origination seems to require a special kind of creative imagination' (*ibid.*). What I find interesting is that Barbour who has a

METHODOLOGY AND EPISTEMOLOGY 131

distinctly theological interest in models should complete his initial discussion of the nature of various scientific models with an acknowledgment identical to that which Black affirmed, namely, that the selection and expansion of models is an activity of creative imagination. We can not pursue at the moment the consequences of such an acknowledgment for our understanding of the nature and construction of scientific theory, beyond observing that the exploration of the place of imagination in human thinking is not confined to the religious field, a conclusion already in evidence throughout our discussion of the philosophers.

From a wide-ranging series of reflections which Barbour has to make on this particular part of our study, I wish to select four for further comment. First of all, he draws attention (*op. cit.*, p.32) to the character of models as analogical, with the implication that in some ways, the similarities between the model and the original are sufficient to constitute a positive analogy, and the dissimilarities to constitute a negative analogy. In simple cases the distinction is obvious. If we say that the 'camel is the ship of the desert', entities such as funnels and humps, or legs and turbines, fall within the negative analogy, while the capacity to carry human beings across trackless wastes (whether of sand or of water) would fall within the positive analogy. Clearly the warning is salutary, for in using any model we have to be careful not to draw deductions from the negative element in the analogy — unless, of course, we are engaged in formal debate, where there is a long-standing practice of doing just that, namely, deducing conclusions from the negative element in the opponent's illustration, or simile, or metaphor, in the hope thereby of embarrassing him. Now Barbour is right to remind us of the analogical character of models, and therefore of the important analogical function of imagination. So far I have refrained somewhat from laying much emphasis upon this side of imaginative activity. Clearly, the literary imagination is constantly engaged in the search for analogies, as it portrays its subjects, by means of metaphors, similes, myths, allegories and parables; as the scientific imagination does in its construction of models. What I would not be altogether happy about, and I am not saying that Barbour would be either, is the application to imaginative activity of a highly sophisticated logical theory of analogy. I say so as one who may in the past have been guilty of an interest in such sophistication. But the mention of the positive and the negative analogy, when it is taken as more than a warning not to draw false deductions from our analogies, when, in fact, it is based on the assumption that we are able to draw a line down the middle of analogies and so separate the negative from the positive components, could be the beginning of the end of analogy. For it could lead to the retention of the positive analogy as univocally and literally true, and to the rejection of the negative analogy as totally false and misleading. Now maybe in some mathematical and scientific models, there is validity in emphasising the negative and the positive analogies. But even there it is not universally relevant. Certainly, when we move into the more literary, philosophical and religious areas of imaginative activity, the distinction could be destructive of the model, if rigorously applied. That is indubitably the signal that comes from our own study of the parables, where we were contending that the similarity or the analogy was contained in the whole parable, and was not to be distilled from it into what could be called the positive analogy of the logion, nor was it to be identified in the manner appropriate to allegory with the parts of the parable.

The second element in Barbour's account of models which I have chosen for further consideration is his account of the status of models. He notes four views of such status: naive realism, or literalism, which regards models as pictorial representations of the world about us, but which is said to be falsified by quantum physics; positivism, which holds models to be taxonomic devices, shorthand summaries of experienced phenomena capable always of being cashed out in empirical statements, a view which does not accord with the exploratory role which models have often played in the advance of science; instrumentalism, which does justice to the prospective role which models play as tools to enable correlation and prediction, but which accords them the standing of mental fictions, or heuristic devices, which bear no relation to reality as in any way representing it; and finally, the view which Barbour himself favours, critical realism, which acknowledges that models are a kind of amalgam of human creativity and

METHODOLOGY AND EPISTEMOLOGY 133

representation of the external world. The title derives from the epistemological theory associated with Kant, and later, with Dawes Hicks, which holds that the world that we 'see' involves elements of external stimulation and subjective interpretation through intuitions and concepts. Perhaps Barbour goes further than is necessary, when (at p.37, *op. cit.*) he avers that 'like the naive realist, the critical realist takes theories' (and by implication, models, I would have to add) 'to be representations of the real world', for as he says a few lines later such representations are not pictures of reality. Perhaps 'representations' is a word to avoid in this area, even when it is associated with the qualifier 'symbolic', which is almost a tautology in this connection, or if not, is to some extent ambiguous. 'Descriptions', which as he says owe much to human imagination and are capable of insight as well as error, might on the whole be a preferable term to apply to models, though it does not do full justice to what the instrumentalists had to say about the exploratory activities which models and theories foster and make possible. At this point the methodological part of our discussion merges into epistemology; and I should therefore prefer to leave for the moment the question of the part which models play in our knowledge of reality, noting however the controlling part which they play in the promotion of any science, as well, as we shall shortly be observing, in the development of theologies.

Before we go on to consider how models, or indeed images also, operate in theologies, we are due to reflect upon a third noteworthy feature of Barbour's discussion of models, namely, his account of how religious models differ from scientific models. This difference is important on two counts. On the one hand, when an extensive examination of the contents of philosophy of science is made the preface to theological enquiry, there is always a danger of construing theological issues and subjects in exclusively scientific-philosophical terms. When I said previously that in theological enquiry, we employ the notion of 'models' as a model, I meant that we were not using it in a wholly literal sense. On the other hand, we have not ourselves sufficiently explored the psychological range within which images and models may operate. The criticism might fairly be made that we have not altogether escaped

Baillie's condemnation from which our whole study started, of presenting our theology in exclusively intellectualist terms. It is just as possible to be intellectualist and conceptualist about imagination, and about the place of imagination in theology as it is in the definition of the *opera ad extra Trinitatis*. Barbour's account of the differences we are considering between religious and theoretical or scientific models is naturally determined by his previous account of the status of models. He agrees with philosophers like Braithwaite and Miles who lay great stress on the capacity of religious models to evoke quite specific religious attitudes, and also to be the inspiration towards, as well as the declared commitment to, a system of noble values and an agapeic way of life. Such are reactions to models which we would not expect to find in the field of science. But such attitudes and such reactions, according to Barbour, have their source in the fact that religious models do provide an understanding of the Reality which or rather whom, they are seeking, however imperfectly to describe. In other words, the difference between religious and scientific models does not lie in the fact that the former seek to achieve the non-cognitive purposes of inspiring attitudes and prescribing moral action, while the latter attempt only cognitive aims; rather do both sorts of model seek cognitive purposes, but the religious models seek to engender attitudes and to promote courses of action and a certain way of life. It is a point to which we should have attached more importance than we did, when considering the place, for example, of images or models in soteriology. The question which we then asked was why it is that in soteriology the Church has retained most of the original biblical models, whereas in most other areas, notably in the doctrine of God, or in that of the person of Jesus Christ, there has been a continuing tendency to introduce more philosophically-oriented models and images. An answer further to those we gave previously (*vide supra*, pp.6off.) could now be entertained, namely, that many of those biblical images carried considerable emotional and even volitional overtones, in addition to the barely cognitive functions which they performed. It is now many years since the late Canon Charles Raven, in a series of lectures on Iona, suggested that God's people have only properly understood and appropriated the meaning of the

METHODOLOGY AND EPISTEMOLOGY 135

death of Christ for them, when it was couched in language and imagery drawn from their own economic setting. To speak of Christ paying the ransom for many by his death was to use language and imagery which came straight out of the social and economic circumstances of the slave population which first heard the message, and to make an immediate emotional rapport with them where they were. For similar reasons, the soteriology which speaks of freeing men and women from bondage not only answers to the condition of our contemporaries in many countries of the world, but also provides them with a volitional base for their actions to throw off the thraldom. A medieval soteriology, moving away from the biblical imagery, spoke of satisfaction offered to the overlord of the universe by the God-man, on behalf of his brothers and sisters who by reason of their sins were unable ever to make good the wrong, by way of disobedience and dishonour that they had done to their Creator. It was imagery that touched them where they had feelings, and it created reactions of gratitude, obedience and commitment. Raven pursued the theme in a variety of directions, but concluded with the proposition that in our time, when we have become so sensitised to the importance of the person and of personal relationships, through the work of Martin Buber and latterly John MacMurray, the image of reconciliation must carry the emotional and the volitional load which other images in other days have been able to sustain, but do so no longer because of social, economic and cultural changes. Barbour agrees with F. Ferré ('Metaphors, Models and Religion', *Soundings*, vol. 51, pp. 342f.) that religious models seem to exert greater influence upon believers than do the doctrines to which they give rise, which seems to bear out the thesis explored by Raven on Iona with these two qualifications: on the one hand, a doctrine such as that of the saving death of Jesus Christ may renew itself in successive generations through the revision of the images in which it expresses itself; and on the other, even images, like the young men, may eventually grow weary, and require to be renewed.

The fourth element in Barbour's treatment of models which shall occupy our attention next is what he calls their capacity to extend theories. He quotes with approval Mary Hesse's argument that 'because of its suggestiveness and open-endedness, a model is a continuing source of plausible hypotheses' (*Models and Analogies in Science*, Sheed and Ward, 1963, Ch. 1). Paying heed to my own warning not to programme an analysis of religious imagination and its use of images and models too literally upon the scientific-philosophical prototype, I should like nevertheless, being guided by this suggestion that a model may be applied to areas of a subject to which it was not originally thought to apply, with a consequent illumination of these areas, to illustrate how effectively this method has been pursued by a selection of prominent theologians of our day. Remembering the emphasis made by Black and then by Barbour on a number of occasions, that the creation or selection of models is a function of imagination, we shall be investigating the place of imagination in theology through the choice of models and images for the explication and exposition of its subject-matter. The examples chosen are purposely wide-ranging, the better and more comprehensively to demonstrate the thesis that theology is image and model centred and controlled.

First, Eichrodt, *Theology of the Old Testament*, opens his first chapter with the affirmation that the concept of 'covenant' binds together the history of the people of Israel, and it unites in one the many components of the Old Testament. The book unfolds the many ways in which the concept achieved both of these ends, and in so doing has provided a normative theology of the Old Testament for many of our contemporaries. Clearly, Eichrodt has not invented this concept, so that we can not say that it is the arbitrary creation of his imagination. On the contrary, it was part of the text which he had in front of him in the scriptures of the Old Testament. Nor could you say that he had argued inductively to the notion by adding texts together and computing their Highest Common Factor; if perchance he did, there is certainly no evidence to suggest that that was how he came to choose this concept to perform the regulatory and expository functions that he had in mind for it. He selected it from a large number of competitors, and gave to it a dominance and a significance which was unparalleled in previous Old Testament theology, because in an act of creative perception he realised that around it could be clustered most of the essential themes of Old Testament theology, and by means of it light could be shed more effectively upon the understanding of the history of the people of Israel in their dealings with God and with their fellows, than by any alternative concept, image or model. He made that original imaginative, creative choice, and the theology of the Old Testament which evolves under the control, domination and illumination of that concept is the evidence for the wisdom of the original choice. The model of the covenant — and here the word 'model' is being used in such a non-technical sense that 'image' would be a fair substitute — with its emphasis upon the inequality of the relationship between a gracious God and his sinful, disobedient people, a God who is under no obligation to love and pardon his people other than that of his infinite love for them, that model is expanded into the formal structures of theology, passing beyond the boundaries of Old Testament theology, to reach something resembling the theories of the scientific method. Once adopted, the model may be used to describe areas of God's relationship with his people where the records might well not have used the term at all. So the pattern begins to emerge of imaginative selection of the model, careful and detailed working out of the theories and doctrines that derive from it, and in return supply the evidence for its validation; and finally, the extrapolation of the model into others of the relationships which God has set up between himself and his people, and to which the model had not been originally applied by the biblical writers.

Our next example is C. H. Dodd's model of 'realised eschatology' (*The Apostolic Preaching and its Development*, 1936) which presents the kingdom as an actuality already realised through the Incarnation of the Word, through his preaching and healing, and through his death and resurrection. This model is a good example of a feature of scientific models to which attention is often drawn, that two components of the model which are not in themselves compatible are put together within one structure. Thus, on the one hand, we have the notion of the *eschaton*, the end-time which is the grand finale of all history and of the entire cosmos, with nothing left beyond that point to happen in history or to the world. On the other hand, we have the notion of realisation, that everything has now been achieved, that everything to which the people of Israel had been looking forward for centuries has happened, and every promise of God made known through the prophets has been fulfilled. Put these two apparent incompatibles together and you have 'realised eschatology', the concept of the end having happened already, and yet of history still happening; of *Christus Victor* who has triumphantly put all principalities and powers beneath his feet, and yet of evil still visibly rampant; of the power of sin being cancelled by the redeeming death of Christ upon the Cross, once and for all, and yet of sinners continuing to sin and to be forgiven, in the process of being saved. Here we have a model for which an original text in the New Testament might be claimed as support, namely, St Luke 17.21, which has been variously translated as 'The Kingdom of God is among you' or 'within you' or 'in your midst', but there is little agreement that this text provides total and incontrovertible validation for Dodd's interpretation of it. Rather does it make much more sense to say that Dodd in an act of imaginative selection, hit upon this model, admittedly after researching his sources very extensively and no doubt having tried and rejected other possibilities. Having selected it, he tried it out on a number of eschatological passages previously held to be cases of consistent eschatology, which with understandable logical consistency regards the *eschaton* as if it really were the end of all things, and concluded that these passages were more amenable to a realised eschatological interpretation than to one provided by consistent eschatology. So he began, and was closely followed by many supporters, to build up his doctrine in a more thoroughly conceptualised way, extending it into areas of theology where it might not appear to be immediately relevant, such as baptism and the Lord's Supper, and more widely to the doctrine of history. Here, then, the contact with the text of Scriptures is not perhaps as strong as it was in Eichrodt's case, though Dodd would not have been prepared to admit it, especially in face of the parables of the Kingdom as he understood them. But even so, we must not confuse the interpretations of New Testament sayings and episodes which are later derived from the model with the grounds on which the original imaginative selection of it was made, for the

METHODOLOGY AND EPISTEMOLOGY

express purpose of the deployment of it in theory and doctrine and in extensive biblical and theological interpretation.

Our next example is not too far removed from the interests and methods of Dodd. Cullmann's earlier form of the 'redemptive line' of history (expounded in *Christ and Time*, ET, SCM Press, London, 1951, but later considerably modifed) is an excellent example of how a theological model is constructed and of how it works. He had observed that the history of Israel as we have it in the Old Testament and the New Testament contains occasions of consummate importance to the narrators as well as to the persons of whom they were writing, what Cullmann calls *kairoi*, episodes in which God displays his grace towards his people, reveals some aspect of his purpose of salvation for them, effects their deliverance from this distress or that, and requires of his people an appropriate response of obedience. The model emerges when Cullmann joins up these *kairoi* to form a redemptive line, to use his name for it, or Christ-line, because it represents a prophetic progress towards, and a further movement beyond, the centre of the line which is the redemptive event of the Incarnation of Jesus Christ. To this centre of the line the *kairoi B.C.* look forward, and the *kairoi A.D.* look back. The line has a beginning before creation, and after the Second Coming, which is also a point on the redemption line, it continues onwards into eternity. This model is very much a picture model, a complex image: Cullmann actually draws it for us. With its aid, he endeavours to destroy the dualism implicit in the *Heilgeschichte* doctrine with its counterpart of secular history, or of *Geschichte* and *Historie*. He produces what might be regarded as a more complete eschatology than Dodd, because he does more justice to the 'not-yet-happenedness' of the *eschaton*, the *parousia*, the Second Coming, and therefore to the equivocal position of Christians caught 'between the times' of the First Coming and the Second Coming, saved and yet in process of being saved, justified but not wholly sanctified, the Church one because it is the Body of Christ and yet visibly fragmented, Christ the Victor and yet evil flaunting its power over peoples and nations. The original model has yielded these extrapolations which spill over into several doctrines other than that of the Christian doctrine of time and eternity. On this occasion, the connection with the basic Scriptures is extremely slender; in fact, Cullmann comes to admit that in effect there is no line as such. He makes his withdrawal on the ground that a biblical model has to be a totally faithful picture of the original. Had he been prepared to say that his model of a line was more of a theoretical model (in Barbour's sense rather than in Black's) than a picture model, then he might have justified his use of it by pointing to the conclusions which it enabled him to reach; though Barbour might then be inclined to say that Cullmann was drifting towards regarding his model as a convenient fiction (this time, a theoretical model in Black's sense), which had no substance in reality. But even that charge would not be altogether fair, insofar as the *kairoi* which Cullmann had initially examined and employed for the purposes not only of drawing his line but also of characterising it as the redemptive line or the Christ-line, were fully biblically-documented. The model was a conceivable elaboration of these occasions, even if it could be misconstrued. The construction of the image of the line was creatively imaginative and it was, and in some ways still is, highly illuminating. Because it comes from outside of the biblical setting, it serves to illustrate two other points in models which have not been apparent in the two previous cases (that is, of covenant and realised eschatology), namely, first, it draws upon the familiar to expound the unfamiliar, as the *kairoi* certainly are, the first time we encounter them in the book, *Christ and Time*; and secondly, it relies upon a certain analogy or affinity between the model and the biblical material which it was designed to describe or even interpret. This positive contribution stands, even though the model has come to be faulted on the infelicity that in order to postulate a line, Cullmann has had to join up, as if they were continuous, points or occasions which in the biblical record were in fact discrete. He even spoke of 'an upward-sloping line' which was worse, and which to his credit he discarded.

So far we have been considering samples of the use of models by biblical commentators and exegetes. Now we proceed to a dogmatic theologian, namely, Karl Barth who as we might expect of a writer so voluminous employs many models, of which, however, we shall select only two. We shall begin with that of the Word of God. This model has an

METHODOLOGY AND EPISTEMOLOGY 141

immediately biblical sound to it, reminding us at once of the Prologue to St John's Gospel. 'In the beginning was the Word and the Word was with God and the Word was God'; or Hebrews 4.12, 'The Word of God is quick and terrible'; yet it is at once evident that Barth, while no doubt drawing the model from such sources, goes on to fit it into a structure different from that which it had in the original. Take for example his well-known account of the three-fold form which the Word of God takes: revealed, written, and proclaimed. Here the Word of God is seen to be distinguishable from Jesus Christ, even though in one of its forms it is identified with him. But this model is elevated to a position of dominance in Barth's theology, particularly in the earlier volumes of the *Church Dogmatics*; and certain consequences are derived from it. For example, great emphasis is laid upon listening to what God has to say; while expectations are that God's communications will be verbalised. If these expectations are unduly heightened, then two serious results follow: first, that if the verbalisation does not occur, then God will be thought to have nothing to communicate; and secondly, that the other forms of divine self-communication are never likely to receive the acknowledgment they merit. For that reason, if for no other — and there are not a few others — visual images and models have not featured prominently in the Protestant tradition, which has been one of hearers rather than seers. The extrapolation is made into the field of ethics where we have the notion which Brunner so widely popularised, of the divine imperative (see the book of that very title), God's word of command, spoken with authority, calling for an immediate response of faithful obedience. So both the selection and the use of the Word of God model were acts of creative imagination on Barth's part. He selected from the many ways in which Christ is described in the New Testament this particular designation, but he modified it from the original occurrences in St John and the Epistle to the Hebrews, in a novel way. He then began to extend it by means of imaginative extrapolations, which proved to be immensely useful and illuminating not only in Christology which might be thought its primary or paradigmatic reference, but also in the theology of preaching and of the sacraments, in the doctrine of Scripture and ethics, as we have just seen. The second model which Barth employs with immense versatility is that of revelation. It is a subject which I have considered elsewhere (*The Shape of Christology*, pp.157-161), and which I shall here relate to our present interest. As was the case with the model of the Word of God, Barth might claim a biblical connection for the model of revelation, in the text Lk 10.22 = Mt 11.27, 'No man knoweth the Father, save the Son, and he to whom the Son will reveal him'. But his imaginative genius does not lie in his selection of this model to be his dominant hermeneutical category, for an extensive use of it is shared by most theologians writing in the middle third of this century, who even went so far sometimes as to employ the term 'Revelation' as an alternative name for Jesus Christ. Barth's uniqueness, the way in which he did demonstrate the role of imagination in theological construction, was his application of the model to different areas of dogmatics. For example, he uses it to expound the doctrine of the Trinity, beginning from the surprising statement that 'we come to the doctrine of the Trinity by no other way than by that of an analysis of the concept of revelation' (*The Doctrine of the Word of God*, p.358). Obviously Barth is not under any delusion that the concept of revelation was the operative category for the Fathers in the third and fourth centuries who were formulating the doctrine; the notion is conspicuously absent from the great periods of doctrinal development. Rather does Barth appear to be offering to rewrite the understanding of the doctrine for a generation which has perhaps moved away from the Aristotelian categories of the Greek and Latin Fathers, and which, on the evidence of its widespread use, will be more at home with talk of revelation. So he begins with what he calls 'the root of the doctrine of the Trinity', the statement that 'God reveals himself as the Lord', and argues to the three-foldness of Revealer, Revelation and Revealedness. Next, when he comes to Christology and applies the revelation model, a number of interesting and quite novel features appear in his presentation. In distinction from Brunner who affirmed that 'Through God alone can God be known' (*The Mediator*, ET, Lutterworth, London, 1937, p.21), and who seemed therefore to be arguing that it is through the divine nature in the two-nature Person,

METHODOLOGY AND EPISTEMOLOGY 143

Jesus Christ, that revelation is effected by God; Barth eventually came to affirm that 'the divine essence expresses and reveals itself wholly in the sphere of the human nature' (*Church Dogmatics*, ET, IV/2, 1958, p.115). If we allow, as we must, that the human nature of Jesus Christ is displayed within the area of the historical events and sayings recorded concerning the man Jesus, then we are obliged to take immensely seriously once again the problem of faith and history, and to face honestly the challenges of historical scepticism that reach us from a long tradition that stretches back, as we saw, to Kierkegaard. It was a problem which Barth never fully faced. But he did use this application of the concept of revelation to the humanity of Christ to good purpose. He spoke openly about the 'humanity of God', being very explicit about a subject to which traditional theology has referred in whispers. He mounted his severest criticism of the protestant orthodox doctrine of double predestination on the base of his conviction that God's purposes for mankind were revealed in totality in the human life, death and resurrection of Jesus Christ; and that, therefore, there could be no place within such a scheme for a *consilium arcanum*, a secret decree according to which God had elected to deal with the reprobates outwith his redemptive action in Jesus Christ. Extending the model of revelation in another direction, Barth used it to describe the activity of the Holy Spirit as 'God's guarantee that revelation will be revelation' — a theme already examined (*vide supra*, pp.61f.), or 'God's making himself sure of us', or 'God from beneath meeting God from above'. To prevent the frustration of his purposes at the very point where they have to be effective, God graciously gives to man the gift of the Holy Spirit so that revelation for man will be what God intended it to be, redemptively transforming. The Holy Spirit is God's guarantee that this end will be achieved; or, put in another way, the gift of the Holy Spirit thus operating within man, as it were from beneath, goes out to meet the God who has come down to earth in the Incarnate Lord. Extended yet farther, the revelation model provides Barth with his anthropology. Since God requires a guarantee that revelation will be revelation, when it is offered to man, man must be incapable of himself appropriating that revelation. In other words, there is no relic of the *imago Dei* in man, no small area in his heart, mind or spirit, where an original untainted righteousness might still be lingering on, to enable him to know God. Gathering these points together, we see how imaginatively Barth has used the model of revelation to enunciate a variety of different doctrines, indicating how central imagination is to the operation of his method.

Somewhat more briefly let us deal with variants of the theme we have been illustrating in Barth. Bultmann, for example (in *Kerygma and Myth*, ET, SPCK, London, 1964), begins from the premise that the images and models of the New Testament, whether cosmological, medical, psychological or anthropological, are all pre-scientific, and constitute an intellectual stumbling-block for modern man, so that the latter can not come to grips with the true *skandalon* of the *kerygma*. Bultmann's word here is as timely now as it was when he first uttered it, with its warning that models and images can impede as well as facilitate the passage of the Gospel. What Bultmann does is to offer us an alternative model, namely, *Existenz*, and all the associated terms and doctrines which existentialists, particularly Heidegger, have constructed like satellites around it: being and non-being, authentic and non-authentic existence, a range of emotions from *angst* to faith, and so on. This alternative model then takes up the themes which had been, unattractively to modern man, stated in the pre-scientific terms of the New Testament, and re-publishes these themes in extension of the new model, in the belief that in this way modern man can not now evade the thrust of the Gospel message, offer and challenge. Here is a very good example of the theorem which we have observed to be developing all the time into a fairly incontrovertible principle concerning the way in which major theologians go about the business of presenting their material. Bultmann, stimulated by a genuinely evangelical desire to make the Gospel comprehensible for our generation, and aware of the very wide range of possible candidates in the philosophical field who might assist him, selects the major categories of Heidegger's existentialism as the model for his exposition as well as for his exegesis. The result is a theology which has proved to be most acceptable to many other theologians and to interested lay folk, for whom it

METHODOLOGY AND EPISTEMOLOGY 145

has constituted a genuine deliverance from outmoded concepts and images, which were impeding full and committed faith. But when he first made that choice, we can not say that Bultmann could possibly have anticipated the success which his programme subsequently enjoyed. It was an uncharted, highly original act of imagination, which continued to operate as he developed doctrinal extensions of it in a variety of directions. In the hands of an expert like Canon John MacQuarrie, this model has been employed for the exposition of major subjects in the fields of doctrine, ethics, and spirituality. But the main difference between Bultmann and the other theologians whom we have just been considering is that his model, *Existenz*, is not derived in any way from the New Testament, but solely from a secular philosopher. That fact in itself does not invalidate such a method of employing models in theology, nor does it amount to a criticism of Bultmann; it would only become so if, as a consequence, less than the full Gospel story were being told, or essential elements of Christian doctrine were undeniably misrepresented or excised from the presentation. Perhaps the classical example of Bultmann's methodology is the employment by the third and fourth century Fathers of the Church, of the Aristotelian metaphysic, in order to create the two-nature Christological model, and so to define orthodox Christology *ex cathedra*.

We shall now round off this review of the place of imagination in the method of certain selected theologians with a shorter notice concerning the imaginative choice by a number of other theologians whose models have given the hall-mark to their writing. Charles Hartshorne, inspired by the metaphysic of A. N. Whitehead, selected the concept of 'process' to be the source of what he had to say about the nature of the universe and of history, about God's being and attributes, about ourselves and our own nature. The use of the model, and the imaginative style of its extension, made belief easier for a large number of people who had found more traditional views of God and nature to be unacceptable. Teilhard de Chardin accomplished a similar end with the integration of what he had to teach about God, Christ, man, history, the universe, evil and so on, around the model of evolution. For Jürgen Moltmann at one stage of his development it was the model of hope that provided the co-ordinates about which he plotted his theological map; for Emil Brunner in ethics, the model of gift and demand; for Joseph Fletcher in his ethical theory, the model of situation, in all its particularity with persons in unique and unrepeatable relations to one another; and for Paul Lehmann in his, the model of a contextual framework for moral relationships. Even in more theoretical fields, for example, in the philosophy of religious language, we have Donald D. Evans operating with the model of self-involvement as the key to the interpretation of such language, not treating it as flat statement but as expression of self-commitment, or the declaration of intent to act in a certain way, or the affirmation of an express attitude to some state of affairs. The late Ian T. Ramsey, who was the pioneer in the application of the notion of models to religious language, was extravagant almost in his own use of models, speaking of disclosure models particularly, leading to the picture of something unknown suddenly and with immediacy dawning upon the observer or the hearer. Attached to this model he had a number of ancillary images which helped it along — the light dawns, the penny drops, the bell rings, and so on — images which in themselves appear to have little to do with one another, and yet when joined to the disclosure model, are seen to be illustrative of it. But other more emotional models are beginning to appear in theology. One we have already noted — the model of liberation, which has become the focal point of a system of theology, which finds that that model brings life and meaning back to old doctrines, endowing them with a dynamism which they have not had for decades. Here we are beginning to glimpse a new feature of the successful model, that it must create a resonance in the minds and hearts and lives of those who adopt it. If I may revert to a topic which was mentioned earlier, ecological concern: one of the great lacks of this lobby is the absence of a model to induce just such resonance, not just in its supporters but in the masses to whom the message is to be directed. For that reason, this concern is not making the impact which its dedicated devotees deserve; nor has the writing which is voluminous been as structured and systematic as it would have been if it had been centred on a strongly imaginative and compelling model. We are still at

METHODOLOGY AND EPISTEMOLOGY 147

the mercy of catch-phrases which will never do duty for models.

I should like now to draw together the ideas which have arisen in the course of this discussion of models, bearing in mind that our interest lies in their relationship to imagination in theological construction. We observe, to begin with, that some of the models are chosen from Scripture, and some are not. That is important, for we can not in the case of the latter claim for them the authority of inspiration, which might have given them the status of revealed truth. Even when the models are derived from Scripture, we have to note that the selection of those few at the price of the rejection of others, and the extended employment and deployment of them into areas for which they were not originally designed, are not scripturally validated. If, then, the models themselves, or at least the theologian's selection of some in particular at the expense of others, are not scripturally derived, how is the selection of models made? The answer that most commends itself, on the suggestion of Black and Barbour, is that we have here an act of creative imagination, the character of which we can fairly describe. The initial choice of the model is not a random, arbitrary act, nor is there anything fanciful about it. It springs from meticulously careful and detailed study of the relevant subject-matter, and from the awareness of the need for some integrating principle to form the basis for the presentation of the theological doctrine or theory. But it is certainly not arrived at by inductive inference from a selected body of scriptural texts. The imaginative activity appears at two other points: first, when the model is applied to the subject-matter, and the latter is expounded in terms of it; and secondly, when the model is extended to promote the exposition of areas of theology to which it was not originally applied. These uses to which models are put by theologians are not designed to serve solely academic ends. On the contrary, as we saw in the cases of Bultmann and of Teilhard, the model was employed as a means of communicating theological doctrines to contemporaries, who were finding some of the older models unacceptable. Therefore, what by this time must be evident is that most, if not all, theologians of stature employ models in their theological construction; and their brilliance lies in the choice of the most comprehensive, comprehensible and extensible model, as well as in the illumination which the employment of the model and the doctrines consequent upon it throw upon the major themes of theology. It is therefore of greatest importance in our study of any theologian, to determine as early as we can what exactly the models are that he is using, and how he is using them. Equally, if we wish to assess the effectiveness of any theologian, we will do well to scrutinise most carefully the models with which he works, investigating whether they are adequate to the load, and whether they lead to any distortion of fundamental doctrines or beliefs. So, too, since as we have seen there is a close correlation between model-choice and model-use on the one hand, and imagination on the other, we have established the central place of imagination in theological method. There is not, nor should there ever be, any embarrassment in this fact. It is not that theologians have suddenly begun to do theology this way; 'twas ever thus — rather like the old man who when told what grammar was, was surprised and delighted to discover that he had been talking it all of his life. Imagination has been part of theology as long as there has been theology.

Epistemology

It remains for us now to indicate the significance of our examination of the place of imagination in faith and theology, for the time-honoured problem of the nature of our knowledge of God. Admittedly, there are those for whom the problem is an artificial one, insofar as, in their opinion, religious knowledge is based upon revelation which has been delivered in completed form to mankind; it remains for us solely to appropriate it. This view which has a *prima facie* attractiveness and piety, and an apparent simplicity, can scarcely be acceptable for two reasons. First, without detracting in any way from the magnificence of revelation, or from what God accomplishes by it in us, we have nevertheless to acknowledge in mankind ways and means of appropriating the heaven-sent revelation. The adequacy of such ways and means is part of the problem of religious knowledge. Secondly, while it is true

METHODOLOGY AND EPISTEMOLOGY 149

that all human knowledge of God has its origin in, and derives its content from, God; still, there are many accounts on the part of men and women, of what is contained in that revelation as it has come to them. That variety of views is another element in the problem of religious knowledge: how is it that one God, revealing himself to mankind, can be so differently known by men and women who do not hesitate to claim for themselves God's exclusive truth? Allowing, then, that in knowing God, they are in contact with a Reality other than themselves, whom they could not know unless that Reality elected to reveal himself to them, how can we hold on to the measure of objectivity implied by that statement, and yet do justice to human participation in the process of appropriation of that objective revelation? Objective divine revelation and subjective human appropriation — these are the two cardinal components in the problem of religious knowledge. From there, let us break down the problem into its several parts. First of all, there is the difficulty of knowing the unfamiliar in terms of the familiar which is all that has previously been experienced, so that the unfamiliarity, the otherness of the unfamiliar, is retained, and not reduced to the familiar. When we come to *talk* of the unfamiliar, we then have the problem of religious language, of how we can employ the language of this world to speak of the other-worldly, without that language going off in the direction either of equivocity or double-talk on the one hand, or on the other, of univocity, with words meaning exactly the same, when applied to God and to ourselves. Religious language has been a central problem of religious philosophy ever since Plato, at least; but it has tended to dominate theology in Britain, in the interests of some, for the past half century. Secondly, a further difficulty encountered in the discussion of religious knowledge is the fact that empirically observable phenomena, events and occurrences, in the ordinary lives of ordinary people, are open apparently to religious and non-religious interpretations at the same time. To these mutually contradictory sets of interpretations, there is no knock-out solution acceptable to both sides. When we remember that such interpretations need not, indeed are not, confined to isolated events, but may extend to the whole lives of people and to sustained periods of history, the problem moves into a discussion of rival systems of metaphysics. Thirdly, often the question of the nature of our knowledge of God is presented as if it were only a matter of how the infinite, eternal and incomprehensible God can be known here and now by finite minds, working within limited time-conditioned categories. But in a Christian doctrine of our knowledge of God, account has to be taken of the fact that the supreme locus for such knowledge is the Incarnation, an event in space and time, and in history, however else we come to characterise it. The historicity of the events involved in the Incarnation, and the questions of how the past can be known in the present, and the present through the past, constitute issues of special importance for the Christian analyst of our knowledge of God. Fourthly, whether we are thinking of knowledge of God which takes place through the reading of Scriptures and the understanding of the life, death and resurrection of Jesus Christ, or on the other hand, through contemporary situations and relationships with other people, it is clear that normally we do not know God in totally unmediated immediacy. Such mediation does not in fact prevent God from drawing near to us to the point of immediacy, so that we are left in no doubt about who it is that makes the offer or presents to us the challenge — a situation covered by John Baillie's now classical phrase, 'mediated immediacy'. The difficulty which arises out of this aspect of our knowledge of God is one of indicating what such media are, and how they are in fact related to the God whom they mediate. A fifth circumstance attaching to knowledge of God is that unlike knowledge of many things in the external world, it is never merely knowledge, but involves other modes of consciousness. Truly to know God is to love him and to seek to obey him. A religious epistemology, therefore, has to be more than that, if it hopes to be that; that is, it must include reference or possible reference to the response of the whole human personality to the gracious approach of God.

Let me now respond to these different aspects of the question of our knowledge, by reference to relevant features of our previous discussions, thus: first of all, on the matter of religious language, the quite special role of parable in the analysis of language must come up for fresh consideration. It is now some thirty years since I. M. Crombie (in *Faith and Logic*, edited by Basil Mitchell, Oxford, 1957, pp.31-83) first gave attention to this possibility in the context of the modern controversy over religious language; but perhaps that very context prevented his examination from taking in the wider perspective obtained from the investigation of the place of the parables in the teaching of Jesus, as pursued by biblical, as distinct from philosophical scholars. Salle TeSelle (McFague) in her two books *Speaking in Parables*, and *Metaphorical Theology* has indeed benefited from such research, as well as from the studies of literary theory in general. Previous generations have perplexed themselves with a whole system of possible accounts of religious language: for example, that the way from the familiarity of ordinary speech to God-talk is by the *via eminentiae*, the extension to ultimate perfection of the highest values we know; or it is by way of negation, the *via negativa*, which negates of God such attributes as we encounter in humanity; or again, as Kierkegaard so often affirmed, by way of the Paradox, which in ordinary logical terms so closely resembles contradiction, but remains inescapable in religious language because of the Incarnation from which all talk of God derives, and which is in itself the supreme Paradox, of the Eternal entering time, God becoming man; or by way of the *via analogica*, which seeks identity-in-difference in human language in its divine reference. How parable 'works', we have already discussed at length, and it is in that direction that we should be looking to understand the nature of both how we talk about God and about how we know him. One temptation we have, I am convinced, to resist, and that is to approach parable as if it were no more than a form of analogy. The parable does not break down into the factors of the positive analogy and the negative analogy, as we have already argued; and it is a danger to which we have equally to be alert, when we choose to equate religious language with metaphor. However, whether we speak as the Bible and the New Testament scholars do of parable, or with the philosophers of religion of metaphor, either way, we are instating imagination at the heart of questions about the nature of our knowledge of God, and about how we speak of the God whom thus we know. That being so, we are thereby committed to including an analysis of imagination in our epistemology.

Turning to the second question mentioned in connection with our knowledge of God, namely, interpretation, we recall that models and images played a decisive role in the construction of theology, in making possible quite specific interpretation of the religious material provided by faith. So there is a sense in which interpretation must be regarded as one of the most important features in our knowledge of God, raising acutely as it does differences of conviction among Christians on particular points of doctrine, and differences between Christians and non-Christians over how definite aspects of history, human existence and society are to be described and understood. Since I have already dealt at length with the subject of interpretation in the theological context (ed. R. W. McKinney, *Creation, Christ and Culture*, T & T Clark, Edinburgh, 1976, pp.216-224), and considered the varieties of interpretation that might be given to interpretation itself, ranging from logical inference, through multiple uses of Wittgenstein's 'seeing-as', to translation, I shall not here repeat what was said there, since I have not had occasion to change these opinions. I would, however, like to make one significant addition: it is, in the light of our previous continuing discussions here on models and images, to emphasise the very important part which they play in almost every form of the interpretation of interpretation, with perhaps the sole exceptions of logical inference and revelation; and consequently, to record the prime place which imagination must be given in any analysis of the interpretative process wherever we encounter it. It is part of the selective process in which the religious person chooses the model in terms of which he or she is going to interpret his or her life and its vicissitudes, or history itself, or the events upon which a religious construction of life, nature and history is going to be based. The direction which this same interpretative process is given and the areas of these subjects to be illumined by such extension are continuing parts of the same imaginative activity. Even when the interpreter claims that the original interpretation and the model or image which gave it its structure and meaning were part of the revelation itself, he or she will have to go beyond the original revelatory situation and extend it in a variety of directions, for which the same confirmatory guarantees do not exist as for the

METHODOLOGY AND EPISTEMOLOGY 153

original; that extension is executed by means of the imagination working here in much the same way as it does in other cases. Once again we encounter the central role which imagination plays in the knowing and understanding process.

The third matter raised in connection with our knowledge of God was that in a Christian context, such knowledge can not be properly discussed out of relation to the events of the Incarnation, and by implication to the biblical comprehension of the story of Israel and its prophets. Previously, the contribution of Collingwood, albeit unwittingly, to the problem of faith and history has been examined (*vide supra*, pp.oo-oo). It would be wrong to suggest that we need not go beyond Collingwood, though it is remarkable that most of the participants in the so-called analytic philosophy of history during the past thirty or so years have used his positions as co-ordinates by which to establish their own. Without uncritically accepting Collingwood's theorems, we may with profit consider their relevance, this time, to religious epistemology. At once I have to say that it is wrong to antithesise historical knowledge of the phenomena recorded in Scripture with here-and-now knowledge of God, as if the possession of the one excludes the other — an antithesis which, as we saw, Kierkegaard came close to setting up. But one way of dealing with such a possible antithesis is to approach the question of knowledge of the past with sympathy for Collingwood's concept of *a priori* imagination. We can then say that in reading the records contained in the Scripture of what God has done in times past in relation to his people, we can really only know these acts and events in the past by means of this quite unique mental activity which Collingwood has called re-enacting the past, by re-thinking the thoughts of the persons described in the Scripture. To achieve this goal — and I would like to suggest that it is one which has been achieved by Christians over the centuries in their study and devotional use of the Scripture: it is no rare phenomenon — we have to enter imaginatively into the situations and events recorded, projecting ourselves empathetically into circumstances and surroundings maybe culturally foreign to us, in fact, in terms of a distinction made previously, even identifying with the persons concerned. So far then is historical knowledge from making it impossible to know the subjects of the faith, that it begins to appear as a necessary and inalienable element in such knowledge — though you might not think so, to judge by many analyses of so-called Christian experience of God. Someone might be inclined to criticise this view by suggesting that siding with Collingwood proves too much, for in his sense, all historical knowledge has the peculiar character of involving us in the past. I would agree, for part of the significance of the Incarnation happening in history is that it falls within the conditions that obtain there, and among such conditions must surely be the conditions under which historical subjects are known and understood. But I would want to add that the historical knowledge which takes place when we read the Scripture differs from other historical knowledge, because of the differences of the persons participating in such history, as well as of the kinds of things that they do and say, particularly of course, God and Jesus Christ. In both cases, it is the epistemological function of imagination to enable us to know the past in the present without confusing it with the present; and, we must add, to know the present in terms of the past. Accordingly, when we speak of knowing God, in Christian terms, true we know him in the present, but also imaginatively in the context of what we know him to have done historically in the life, death and resurrection of Jesus Christ, showing us there the vast dimensions of his love and forgiveness, the tenderness of his compassion towards us and his care for us. The religious imagination holds that past in this present, and the two together form what we call 'knowing God'. Such a view does not entail that we accept as literally true everything that the documents say, in the way in which they say it. In Collingwood's own case, the theory of the *a priori* imagination was combined with the most rigorous criticism and analysis of the relevant historical records and evidence. In fact, quite a large part of his study of the nature of historiography is devoted to demonstrating that he does not accept authorities as such; he selects, and allows them to modify one another. After all, in I Cor 14, St Paul speaks of testing the spirits and judging the prophets. Collingwood, therefore, writes what he calls 'critical history'. In the same way, we should expect the scriptural records, through which knowledge of God is mediated to us, to be open to the normal

METHODOLOGY AND EPISTEMOLOGY 155

processes of criticism, and more, that the historical imagination might well be employed in the assessment of different sections of the text. That is part of the way you think historically-critically. To sum up, then: the importance of imagination for the understanding of our knowledge of God in a historical context, as we believe is the case in Christianity, is two-fold. First, it is present in the process of critically assessing the records which come to us as the historical foundation for the Christian faith, as it is in all historical knowledge. Secondly, it is involved in our knowledge of the past thus recorded, so that though past it is known only through our re-enacting it now, and through our re-living of what is recorded for us.

The fourth circumstance which we observed in our initial analysis of some of the problems and peculiar features of our knowledge of God was its character of 'mediated immediacy'. It is a phrase which may be variously illustrated, for example, as we have seen in revelation, where God is known through that which is other than God, say, through the human nature of Jesus Christ; or again, as in the parable in which Christ explains that the Lord is fed, given drink, visited, and clothed, insofar as these kindnesses are shown to his brothers and sisters who are distressed in any way. But, in the light of our discussion of models, images and parables, I would want to add that they play a highly important part in the process of mediated immediacy, for in many circumstances they constitute the media element in the complexity of that process. Undue concern with the question of how far a given model or image may be representative of its original, leading to the enumeration of the negative and the positive parts of the analogy, obscures the other way of viewing the relationship; that is, to say that the model or the image is the means whereby we apprehend or understand the original. In other words, they belong, not so much to what is known, as to that by which the object, or rather, the Subject, is known. Clearly, we are not dismissing the possibility of there being similarity between the model or image and the original to be known — indeed, there often will be such similarity, as when the image of fatherhood is used to describe God; but the suggestion is that we put the matter another way, and say that we know

God in terms of fatherhood. This suggestion can be extended even to the parables, to give the theorem that we understand the subjects to which they refer, in terms of them. The Kingdom of God is understood by us in terms of the plethora of images to which Jesus likens the Kingdom, which seems to me to be exactly how Jesus presents them. We understand Jesus' own conception of his redemptive presence and mission on earth, by means of the parable of the wicked husbandmen (Mt 21.33ff.). I repeat, such a way of looking at our knowledge is not intended to deny the similarity between the image and the original, for clearly some images, models and parables will be appropriate and others will not; but also, the supreme control in all such knowledge is the Subject known, and the determination of appropriateness of the models, images and parables lies with him. Sometimes, too, this control will operate through the correction of the models, images and parables by one another, or, as so often happens, by their supplementing and complementing one another, for the ultimate truth is never solely contained in any one by itself. Another way of putting this same position is to say that the models, and so on, are the interpretative structures which we employ in the understanding of religious subjects. Such a recognition brings interpretation into the centre of epistemological discussion, which is then conceived of in broader terms than that merely of the analysis of how we describe what we know.

The fifth feature of religious knowledge which was mentioned earlier was the fact that it is, in a sense, an abstract from a more complex situation than one only of knowing, for knowledge of God occurs within the complex of faith, which, in its turn, involves affection and volition, as well as knowing. I am not anxious to develop that investigation farther, but rather to use its allusion to the modes of consciousness, as an older psychology called them, to draw attention to a major omission from analyses of the place of different psychological activities in the process of religious knowledge. Despite the strong emotional considerations mentioned at the beginning of our study, which accounted somewhat for the ostracising of imagination from the religious scene, it still remains incredible that in epistemology, where such emotional elements could be thought to be properly set in perspective, the ban should be continued. When it is realised that almost every other mental activity has, at one time or another, been included in the Christian account of the relationship of the human soul to God: knowledge, acknowledgment, understanding, apprehension, experience, belief, and conviction — all within the cognitive mode of consciousness — not to mention faith, trust, hope, love, adoration, obedience, commitment, prayer (in all its forms), and worship — all in a combination of the modes of consciousness, though all dependent upon the cognitive mode; we begin to realise how incomplete our epistemology must be, as a result of that omission. Such an omission is not altogether incomprehensible, as we recall how erratic and fitful has been the analysis of imagination in the field of philosophical epistemology; there primacy of interest has centred upon sense-perception and reason, so that even when it has entered into the thought of a major philosopher like Hume or Kant, its place there has all too often been ignored. In fact, so pervasive is imagination throughout most of the mental activities noted above, that not only is it artificial and contrived to omit imagination from the analysis of them, but more importantly, its omission leads to a misrepresentation of their nature. Put positively, the recognition of the place of imagination in these religious and theological enterprises, will draw attention to the openness, spontaneity, variety and originality, which one has to admit are not derivable from ancient formulae and case-hardened argumentation so often used in their analysis. Therefore, it is no longer a question of whether we are prepared to include imagination in our examination of religious knowledge, as part of the whole faith-situation. It is already deeply involved in the whole of it. The only question that remains is whether we are willing to make our analysis of faith and knowledge, and all the activities involved in our relationship to God, complete by acknowledging the pervasive part that imagination plays in them.

Chapter 7

AN ANALYTIC OF IMAGINATION AND IMAGES

Imagination

While it may have been the practice in medieval philosophy to regard imagination as an isolable faculty of the mind, and while even a writer as circumspect as Mary Warnock refers to it as a 'power of the mind'; nevertheless, what has now emerged is that the imagination is the whole mind working in certain ways, which we shall in due course itemise. In much the same way, I would not regard 'reason', as I have said, as a separate faculty of the mind, but rather as the whole mind working in identifiable ways — arguing from general principles to specific conclusions, or from collections of cases to universal judgments, or assessing the relevance of alleged evidence, or the validity of conclusions, and so on. Therefore, on the basis of the different areas in which in the course of our study we have encountered imagination at work in religion and theology, in faith and in moral activity, let us now summarise our results in an analytic of the different characteristics of, first, imagination, and then, images. These items, though understandably similar, nevertheless are distinguishable.

One, imagination is sensitive to, and *perceptive* of, features in the world and in persons which the ordinary observer passes unnoticed. Artistic imagination is able to detect in a landscape, in the sky or on the sea, colours we only begin to notice when our attention is drawn to them, or the portrait painter penetrates to the character of his sitter with insights which are denied to the rest of us. But such perceptiveness is an important aspect of the religious imagination, as we noticed in

Jesus' relationships to persons in many New Testament episodes, and in the Christian ethic which regards as a virtue, sensitivity to human need and suffering in those around us. When we find this feature of imagination in the field of knowledge of God, we describe it as openness to his activities, in history, in his providential over-ruling of the affairs of men and women, or in the details of our own private lives; and as openness to the out-pouring of his Spirit upon our spirit, so that we share in his imaginativeness.

Two, imagination is *selective* from the mass of material with which the mind is ordinarily confronted, and from among which it concentrates upon the salient and we might add, the significant features. Over and over again in the Old Testament, the prophets seized upon this occasion or that in the past history of the nation, out of all the events that had befallen God's people, to demonstrate that God was working out his promised purpose of their salvation. From the literature of that same Israel, Jesus selected such themes as the redemptive ransom, the concept of the Kingdom, or the Suffering Servant, to be the load-bearing topics of his *kerygma*. In Jesus' hands, such selection was also correction, as he indicated from time to time with the words, 'it hath been said unto you of old, but I say unto you ....' Therefore, at any time, imaginative thinking has to take something, and leave something; and always it operates at the risk of distortion or caricature. At its best, it gets to the root of the matter, in much the same way as a portrait is able to do what no photographic copy ever can.

Three, because it is selective of the significant features of a situation, or a piece of history or of literature, imagination is also *synoptic and integrative*. Having selected the salient and load-bearing feature, it proceeds to arrange and systematise the material around it and in terms of it. In the strictly aesthetic sphere, Kant referred to 'purposiveness without design', thus drawing attention to the internal organisation and unity of the work of art, which exists from within, *a se* as medieval theology would say, and is not dictated by some extraneous purpose which the artist had in mind in creating it. The synoptic and integrative roles of imagination were in evidence, in a rather different way, in the writing of a good deal of so-called 'biblical theology', in which some dominant concept or image was chosen by the theologian, and then used to draw together into systematic unity a wide range of different passages and themes. But we also saw that the practice was followed extensively by systematic theologians, such as Barth, Tillich, and Hartshorne, not to mention Schleiermacher and McLeod Campbell.

Four, such systematising and integrating we must not regard as a kind of intellectual cabinet-making or joinery. It requires a high level of *creative and constructive* thought to put together the diverse elements of a story or pieces of literature into unitary form. It is not creative in the sense that it takes off from substantial material into flights of fancy. It has its feet very much on the ground, and it takes reality, the text or history very much as its starting-point. With a writer like Barth or Tillich, the dimension of the structure thus creatively constructed is monumental; in Tillich's case, it has even been said to be monolithic. But the same character appears in the scientific constructions of Einstein, or the literary creations of Tolkien, who oddly enough creates a prior set of documents in pseudo-ancient script, which are alleged to be the source of his extremely systematised epic narrative.

Five, another characteristic which we can not omit from explicit mention, though it is implicit in all that has gone before, is the interpretative capacity of imagination. Though it is impossible to give any simple logical account of the nature of interpretation, nevertheless it is inalienable from any situation in which the human intellect understands any subject, and endeavours to explain it to itself or to others. The interpretative role is fulfilled through the imagination observing analogical connections between entities which less perceptive intellects miss. It is the perception and the extrapolation of such analogical connections which provide the framework for the process of systematising, mentioned under item three above. But it is no less part of the enrichment of our experience when the poet interprets one area of that experience in terms of, and through the images of, another part.

Six, at this point we must have the courage of our convictions and assert that imagination has a *cognitive* role to play in our intellectual lives. In other words, there are things that we would not know about the world around us, about other persons, about our obligations, about ourselves, about the Bible, about history, about doctrine and about God, had we, or as is more often the case, had those into whose labours we are entered, not employed the imagination. These things that we would not otherwise know are not phantasies, illusions, delusions, or hallucinations. These are facts about the real world which would remain otherwise beyond our ken. The point was made in a traditionally philosophical way by Hume and Kant; but Mary Warnock took it up to make it the theme of her book *Imagination*, that the imagination which works so creatively in poetry, painting, music, drama and all the arts, is that same imagination which we employ in our perception of the world about and the people in it, the same imagination functioning in a heightened way and in a variety of different media. Further, if imagination has thus a genuinely cognitive role, we have to revise our idea that it adds characteristics to reality which do not exist at all, or exist 'only in the mind of the beholder'. On the contrary, it is aware of dimensions of reality which are hidden from the unimaginative. In effect, reality is multi-dimensional and richly complex, and in our knowledge of it imagination plays a clearly definable part, one in fact without which much of reality would remain unknown.

Seven, up to this point we might be in danger of giving too intellectualist an account of imagination, and so we must at once draw attention to its *empathetic* function, the way in which it is able to project us not only intellectually into deeper understanding of the situation, but also affectively and emotionally into it, so that we identify with its components and with the persons involved in it. This characteristic of imagination is important in a number of ways. It enables us to feel for the plight of many people in suffering and distress, conditions which are not even comprehended if our approach is purely intellectual. Such empathy, or identification, with them will then form the basis of decision to relieve their plight. It is part of this same function of imagination to devise ways in which others may participate in the experience which it has itself appropriated. It has to make publicly accessible, largely as we shall notice by means of a range of compelling visual or auditory imagery, what has begun as an intensely personal and subjective experience.

Eight, at this point we are beginning to pass over to another equally important function of the imagination, namely, its *communicative* role. It is the responsibility of the artist, the poet, the dramatist to make us feel as they do about a certain subject; and to achieve this end they must not only *have* the experience; they must also create the media for their interpretation and appreciation of it, media which place us where they are, enabling us to see with their eyes and hear with their ears. The difference between the genius and the commonplace lies at the point of communicative capacity. The one has it; the other does not. I wonder if I may go farther and say not only that communication is a role of imagination but also that imagination is the only means of effective communication. How often have we said about a dull play, a dreary film, or an unexciting painting, that 'it says nothing to us', that nothing is coming across from the artist but his dullness, and more, that he is failing to communicate what the situation, the event or the character is, which or whom he has experienced and wishes to communicate to us, in order that we may feel as he does about the subject. He loses us somewhere along the line, and that is the explanation of the dullness. It is a point which no preacher dare ever forget; and the point has a particular relevance for religious educators, who in this generation are very exercised over the question of how they can communicate to the next generation the value-appreciation as well as the religious commitment which they themselves have, without inducing the boredom associated with indoctrination. The moral of our present tale is that communication is supremely, if not solely, by imagination.

Nine, there are two other functions of imagination, which relate to time, the past and the present, which may now engage us. The first arises in relation to history, and to Christian faith which is so firmly based upon history. We might call both of these the *contemporanising* function of imagination, and the first has to do with treating the past as present. It is found first in knowledge of history, which as Collingwood argued involves re-living the past, and re-enacting it in thought. Such re-enactment takes place in the present, so that Collingwood's *a priori* imagination deals with the past as present. So, when we come to Christian faith, we note again the notion of the 'contemporary disciple' who, though he or she lives two thousand years after Christ, can regard himself or herself as his contemporary, so that when Jesus speaks to the rich young ruler, or to the Pharisees or Peter, it is as if he speaks to this contemporary disciple. Present-day believers who read the New Testament with its narratives of the past, imaginatively read that past as present. When these narratives include also the accounts of the resurrection of Jesus, then they will be reading of the One who was dead and yet lives, who died and was raised again and is now present. Imagination could therefore properly be said to be the medium through which the resurrection occurring as it did so long ago is to be regarded as happening again through the category of the past-as-present. In this context, St Paul's statements about our being raised in Christ are given fresh connotation.

But there is a second contemporanising to be found in the New Testament, that of the future with the present. Realised eschatology, which we examined earlier, is a good example of it, the idea that the Kingdom which in one sense is still future is now present, extended as we saw and applied to baptism, the eucharist, redemption and ethics. Some theologians have interpreted this notion in terms of the fore-shortening of the time-span between now and the end-time; others have by contrast dropped the futurist element altogether. Some have even used the imagination's capacity for vivid presentation to say that it views the future so realistically that it believes that it genuinely is the present. Personally, I do not believe that it is necessary to employ any of these devices; in fact, in one way or another, they all falsify what happens with imagination. It will present the future vividly here-and-now, but it will not cease to believe that it is the future. It presents the future-as-present, fully aware that there has been no absorption of the one by the other, and that consequently there is the unfulfilled span of the future still remaining. We may now draw attention to another form of the contemporanising function of imagination, namely, the way in which in the process of decision-making it presents two alternative courses of action as if they

IMAGINATION AND IMAGES 165

were actually occurring, and sees them as actions in the lives of two rather different kinds of person, the kinds of person that we would become if we were to act in either of the ways projected. This contemporanising form of imagination has therefore a large anticipatory element in it, which is of importance in all moral activity, in that it enables us to reinforce one line of action rather than another, even ahead of our being required to do it.

Ten, a not greatly dissimilar function of imagination can be defined in relation to space, which I call its *conspatialising* function. The role which the imagination here plays is to make the absent present. Sartre, it will be remembered, referred often to the way in which a portrait or a caricature will make an absent friend present to me; and he added that it is the function of imagination to do that also. But what can be done in the case of my friend may also be done for peoples of other nations, and other classes, and of much worse economic and political circumstances than ourselves. They may be five thousand miles away, but imagination treats the absent as present, in our midst, with claims upon us as immediate as if they had already knocked on our door. We have come to refer nowadays to 'the global village', as if this were some new product of the technological age or of the world economic crisis. In fact, for the Christian the world has always been a global village, our neighbours however distantly absent they are, being in fact the folks next door. What the Psalmist said in Psalm 139.7-10 about God, that wherever we go, ascending into heaven, making our bed in hell, flying on the wings of the morning and dwelling in the uttermost parts of the sea, there we should still be in God's presence, is also exactly true of our neighbour: he is ever to be regarded as present in absence.

Eleven, there is a remark of Iris Murdoch that we may explore, namely, that imagination creates what she calls 'our-world', which we constitute by our system of values, our principles, but also by our prejudices, as well, we might add, by our religious commitments and subjects of faith, wherein appear all our hopes, ideals and ambitions, and also, if we are at all honest, our phantasies and delusions, not least of all about ourselves. Some of that world we obviously create for ourselves; some of it emerges from the pressures upon us; some we will wish to say comes through the grace of God; but on the human side the structure is formed and the tone set by our imagination. It is the framework within which our decisions are made, our ambitions defined, our emotional reactions stimulated, and in most general terms our lives are lived.

Twelve, I should like finally to nominate the *ecumenical* role which the imagination may play, in inter-church discussions. To explain what must appear to be an unfamiliar place in which to look for, let alone find, imagination, I shall have to provide an enlarged context for my claim, requesting pardon for the introduction of personal details. During a visit to the Gregorian University, I was privileged to have conversations with Professor G. O'Collins and his colleagues on a paper previously prepared on the subject of 'Scripture, Authority and Tradition'. It was presented in a reduced form at the so-called 'Hand-in-hand' conference of representatives of the Scottish churches in Iona in June 1984, and published later in the Catholic ecumenical review, *One in Christ*, under the original title. In the first half of this paper, I had been tracing parallel themes in Catholic and Reformed thinking, for example, about Scripture and tradition, pointing out how in both systems, there is to be found belief in the inerrancy and infallibility of Scripture due to its having been dictated by the Holy Spirit, a circumstance which imparts to Scripture its unique authority; in both systems an acknowledgment of the part played by 'the grace of God and the interior help of the Holy Spirit' — to use phrases from *The Documents of Vatican II* (p.118) which echo almost to the letter Calvin's own words about the *testimonium internum Spiritus Sancti* to the true interpretation of Scripture; in both systems, also, the presence of tradition, overtly affirmed in Catholic theology where the Church through its teaching office or *magisterium* interprets what has been handed down from the Apostles and hands it on (*tradit*) as the growing deposit of faith, but equally visible in the Reformed Churches, who in their Confessions, the *Augsburg Confession*, the *Helvetic Confession*, the *Scots Confession*, and the *Westminster Confession*, to mention only a few, set up their interpretative mechanisms to act as guidelines to the understanding of Scripture and the faith; while the ultimacy of the Church through the *magisterium* as the determining voice in both the interpretation of Scripture and the understanding of the faith is not altogether different from the statement found in the Preamble to the questions to be put to an ordinand of the Church of Scotland, that the Church is itself to be the sole judge of the agreement of any interpretation of articles in the *Westminster Confession* with the Scripture.

This contextual rehearsal ends with the point that is of great importance to our notice that imagination has an ecumenical role to play. Both systems acknowledge that to enable its contemporaries in any generation to understand and to appropriate the faith, the Church has to *interpret* the faith once for all delivered to the saints; and as has been argued extensively above, imagination lies at the very heart of the interpretative process. Let me take two examples, to illustrate how close the two systems are to one another, in their use of the imaginative category. A Protestant, David H. Kelsey (*The Uses of Scripture in Doctrine*, Fortress Press, Philadelphia, 1975, p.163) writes, 'At the root of (any) theological position there is an *imaginative act* in which a theologian tries to catch up in a single metaphorical judgment (or model, as we were calling it) the full complexity of God's presence in, through and over-against the activities comprising the church's common life, and which in turn provides the *discrimen* against which the theology criticises the church's current forms of speech and life'. This imaginative act is a construal of how God is present and acts in the ongoing existence of the church; and it entails the integration into a single characterisation of a highly complex body of material both past and present. He goes on to apply this model as the key to the understanding of a wide range of Reformed theologians from B. B. Warfield to Karl Barth; and to extend it into doctrinal, homiletical and liturgical modes of discourse. A position very similar to Kelsey's is expressed by George Tavard (in an essay in *Scripture and Tradition*) who, pointing out that the tradition does not exist *in se*, but comes into existence 'in the hermeneutical process by which the theologian, criticising the testimonies of the Christian past, assessing their value for the future' (p.97) in an act of imagination (p.121) selects that tradition from the past which, when it has passed through the crucible of his reflection and interpretation, will best enable understanding of the faith by his contemporaries.

The signal which I receive from this all too brief comparative study of how two theologians themselves express their method of thinking and writing theology is that in most ecumenical conversations we waste too much of each other's time upon examining the final products, the end-terms of our theological thinking, particularly about Church, ministry and sacraments. At that point our thinking has set, almost the cynic might suggest with the rigid inflexibility of concrete. On the contrary, it might appear that our only hope in ecumenical conversation is to break into the process at a much earlier stage, when the whole of the material is in a much more malleable condition, when the imagination is still operating, selecting appropriate models and applying them or extending them, deciding which tradition is most worthy and appropriate for extension into a modern setting. We might then be able to compare the models and features from which we have respectively taken our departure, to assess them, criticise or defend them — all at a time when we can better understand how we have respectively reached our conclusions, and perhaps even hope to approximate more closely to one another's positions. Ecumenical conversation has been too often about arriving, about 'Here stand I; I can no other', and much too little about travelling hopefully, when we can still influence the directions we are severally taking. It is not too much to hope for, that we examine more than we do the thinking and the imaginative selection and construction which go into theology, the choice of images and models and the way they are extended and extrapolated; and pay less attention to the results of such processes, in regard to which we are so often obliged to make the choice of accept or reject, the choice of total absorption or mutually exclusive, if also courteously polite, pluralism.

Images

The analytic of images will, not surprisingly, follow a similar pattern to that of the analytic of imagination, but it merits a separate treatment, insofar as there has grown up a body of thought about the role that images play in knowledge,

IMAGINATION AND IMAGES

communication, emotion and even logic. Moreover, it is necessary to reflect upon what exactly it is that imagination uses when it executes the functions which were described in the previous analytic. Let us then itemise the particulars as before:

One, we begin with the *epistemological* role of images which we have encountered in several connections according to the kind of connotation we give to the term 'model'. When it included models of the kind examined by Black and Barbour, the epistemological function was high; for not only did they clearly facilitate the description of entities of a scientific sort, but in the case of many of these, as also of subjects known by means of models in other spheres, they were the only means by which they could be known. We might here cite Tillich's statement that there is no non-analogical knowledge. But without going as far as Tillich, we observe that the images we encounter in the Bible — of God as a rock, a strong tower, a shepherd, and most significantly of all as Father; or of the death of Christ as ransom, propitiation, sacrifice and redemption — are descriptive of God and of the cross, and are the subjects of knowledge. So close do they stand to the subject of knowledge, that we have no ground for saying that these images somehow represent a reality which in itself is ineffable, or transcends description. We may wish to say that some of these terms are used analogically when applied to God or to the death of Christ; but even so, the positive content in their analogy is sufficiently high to justify our claim that in knowing God thus in terms of these images or models, we have genuine knowledge and are not being deceived. When we turn to the parables of the New Testament which we considered to be extended images, the same claim will be made that to know God's forgivingness as it is explained to us in, say, the parable of the Prodigal Son, is genuine knowledge, even, some suggest, the most effective and accurate knowledge possible for us.

Two, another way of describing the epistemological role of images is to say that they are mediative, the media whereby we know certain subjects of the faith and of theology. When their status is thus defined, while we do not rule out the possibility that in some cases there may be similarity between the image or model and the original, an analogical relationship which might even approximate to representation, nevertheless we are not obliged to hold that the relationship is that of one-to-one representation. The device which we are here employing to define the status of images and models is akin to the way in which the question of the status of the so-called 'primary and secondary qualities' was answered in ordinary epistemology. Over against the doctrine of representative perception which held that such qualities, the first of which actually reside in objects, and the second are produced in our minds by the primary, represent to us otherwise unknowable objects, a more acceptable theory was put forward, namely, that we apprehend objects in the world about us in terms of such qualities, or more exactly, in terms of the sense-data, which they stimulate in us.

Three, in a now famous statement, Paul Tillich once wrote that symbols participate in the reality which they symbolise, care being taken to differentiate signs and symbols in this respect — fatherhood in the reality of the God who is father of all, sonship in the reality of Christ who is Son to the Father, reconciliation in the reality of the death of Christ, and so on. So understood, symbols, images, and models acquire an *ontological* role and a status within reality. They form part of our world, and when they integrate with value-systems, then their metaphysical character becomes clear. When the symbols of any culture collapse and its world as a result disintegrates, then that culture is in rapid decay. That is what is being referred to when the social analyst says that contemporary society requires a new set of symbols to give or restore significance to daily existence. Such symbols as part of the ambient reality act as direction-finders, trig points for decision-making, for choices between this course and that. But the concept has a religious significance as well. One important application assists us in understanding the theology of the sacraments, and particularly one very controversial feature of that theology, namely, the relation of the bread and wine to the body and blood of Our Lord. On the present reading of the nature of images and models, the bread and the wine participate in the reality of the body and blood of Christ, without thereby standing in a purely external relation to them, such as representing them or being signs of Christ's presence, or, on the other hand, being so identified with them as to lose their creaturely attributes, and be metaphysically transformed not just into the substance but into the substance and attributes of the body and blood of Jesus Christ. This view can then be combined with that which we stated earlier (*vide supra*, pp.35-39) about the imaginative proximity of bread and wine to body and blood of Jesus Christ. So, too, an image like reconciliation, once related to Christ through his death, carries henceforth that content; it is re-constituted, and now is to be understood in that primary sense.

Four, thus construed the image will inevitably develop a *hermeneutic* role and become the means of interpreting sections of Scripture, or areas of theology other than that of its original setting. Cullmann, it will be recalled, used the image of *kairos* as so often did Tillich to describe situations which were not originally included in the concept. Teilhard extends the image of evolution away beyond its original application. The interpretation of interpretation which I favour is that it consists of transposing a concept, sentence or passage, from one culture-system into another, so that the new formulation has the same meaning within the new system as the concept, sentence or passage had within the old. Now the effective agent in such a transposition is the model or image which is selected in the new system as the bearer of interpretation. Sometimes such an image or model is too far-fetched, as when, say, maladjustment is employed as the image to interpret sin, in a sociological context. The choice of adequate images, symbols or models by the imagination is of paramount importance in the translation of ancient doctrinal truth into the language of modern times.

Five, as a direct consequence of the hermeneutic role of the image comes its *constructive* role, as it is seen to be the means whereby systematic structures of theology are set up — Eichrodt employing the model of the 'covenant' to build the integrated fabric of a theology of the Old Testament, Ian Ramsey using the image of 'disclosure' to construct a philosophy of religious language, as D. D. Evans did with his image of 'onlooks', or John Hicks with his symbol of 'seeing-as'; Paul Lehmann constructing an ethic on the image of 'context', and Joseph Fletcher on the symbol of 'situation'. Using our own model, we might say that the image is the cement of several systems, for it is through it that their component parts cohere, and their compresence within the total system is understood.

Six, the image has a very interesting further function, namely that of *universalising* an experience or event which had a subjective character, or one which was very much bounded by its original historical origins. The essence of good art is that the artist should discover that image, that form, which will enable him by publicising his experience, to communicate to others what had up to that point been purely private. The purpose of this expression is that others should come to re-enact what was his experience, to relive it, and know it not just by description but from the inside and by acquaintance, to use the old distinction of Bertrand Russell's. That is the goal which is aimed at in some forms of religious education, the choice of the right symbols and images, to enable the next generation to participate in the experience of the faith which their elders in their day had received in the same way. It is the peculiar quality of the image at once to be particular, specific and private, and also to have the universalising function to which I refer, that of lifting an experience, or an aspect of the faith, out of the immediate consciousness of one person, and placing it at the disposal of all and sundry. The same responsibility and opportunity await the preacher who takes his message from some passage in the Bible, and wishes to remove it from the particularity of its original occurrence in biblical history, and to place it at some point of public availability, where his generation will hear the story in its own accents. Another way to describe this role of the image is to call it *communicational* or *kerygmatic*, for it is an essential feature in the communication of the Gospel, and in the discovery of the most effective means for the transmission of the Gospel.

Seven, partly in the communicational context, but also in others, the image may be said to have an *illuminative* role. In theological study, an image may suddenly throw light on other areas than those for which it was chosen or designed. The two-nature theory in Christology found its way into the doctrines of Scripture, the Church and the sacraments; while 'crisis' was used to explain a whole range of theological subjects apart from the Last Judgement. So we might speak of the capacity of a good image or symbol to be deployed to include subjects which it was not planned originally to describe.

Eight, sometimes this illuminative role is extended to the point where it hardens to become *regulative* and *prescriptive*. The image which was adopted for one part of the discipline acquires some kind of authority which is thought to empower it to prescribe the contents of other parts. I can imagine someone who has become, we might almost say, addicted to existentialist categories applying them to areas of theology for which they had not been specially selected originally, or considering them to have been so successful within a limited scope, allows them the run of the discipline, prescribing and legislating for details well beyond their original bailiewick.

Nine, it is then a short step to endowing the image with a *normative* character so that it becomes the criterion by which truth and falsity in theological statements are assessed. The reference to existentialist images is not out of place in this connection, for not infrequently we hear of certain theological views being judged by the possibility of their convertibility into existentialist types of statement. But the practice is not confined to existentialists; we all do it. When we become involved in theological controversy, the standards by which we tend to offer criticism of views with which we disagree are composed largely of our favourite images and models. That is why so much theological controversy is so inconclusive. We and our opponents both think that we are arguing on common ground about certain theological assertions, whereas all the time what is at issue are the images and models, often undisclosed and maybe even unconsciously adopted, which govern most of our thought.

Ten, we have, therefore, to add to our growing list, that images have a logical function, which is both *methodic* and *argumental*. They are methodic in the sense that they are central to the method that we each of us employ in the construction of our theological views — even when we disclaim that we ever construct theological views, adding that we do not aspire to do other than understand them. Allowing for such proper modesty, we have to point out that understanding involves interpretation; and that is a process which is heavily laden with images of one kind or another. So, too, when we endeavour to argue our way through some theological issue, we shall find once again that we proceed with the use of images in the form of controlling categories and concepts, some of which will have clear visual connections, many with parabolic overtones and others with the modular character of classical theology. These images, if I may use the inclusive term, stand at the great turning-points of our arguments, determining where we go next, and how the argument may be best stated.

Eleven, it would be remiss of us to omit the singularly *evocative* role which images play. A great deal is made of this aspect of images in the arts: the use of a vividly pictorial phrase, a sharply sketched scene, even the phonic quality of the spoken word, will serve to induce some emotion or other. In the visual arts, often the whole purpose of the imagery is, not to do any of the previous ten things we have so far been speaking about, but purely to evoke in us the feeling the artist himself had in the presence of the object portrayed. We do not require to go far before we encounter massive evidence for the evocative role of religious images — in crucifixes, ikons, religious paintings and sculptures, in the whole architecture of a church, as well as in the acted imagery of the entire liturgy. It is all designed, under and within the glory of God, to place us at the point where God will meet us and we him, and be renewed through this encounter with him. For that reason it is quite unjustified to condemn as formality a ritual which you do not understand, because, for the practitioners of that ritual, it may be the means by which there is evoked in them response to God. For them it may well be the means whereby the God whom they had thought of as completely absent is recognised as being present-in-absence.

Twelve, the image must also be said then to have a *sustentative* role, insofar as it will often be the means whereby a flagging faith is sustained, a weak will is reinforced, and a fading conviction restored to full strength. No matter our denominational loyalty, we all depend heavily on the sustentative power of images, not because they have any power in themselves, but because they have been in our lives the means whereby God has so constantly refreshed and renewed us.

IMAGINATION AND IMAGES

Thirteen, a final role of images is their *recreative* character. I am thinking first in general terms of the way in which the sight of a keepsake given by an almost forgotten friend of other times and other places, may give us a very immediate experience of him or of her as present-in-absence. The reading of an old letter will often revive emotions which we had long thought to be dead. The words of an old hymn learnt in childhood may bring an unrepentant sinner to himself and send him on his way seeking forgiveness. There are many examples of this aspect of images in popular religious literature, but we dare not despise it. It is the stuff of which penitence is made, and made much more quickly and effectively than by some of the heavy hell-fire incantations that are sometimes thought to be the proper approach to unrepentance.

So, in conclusion, the question is very often asked: how can we ensure that the role of imagination in religion and theology is not taken over by phantasy, and how, therefore, can we distinguish them from one another? It might seem that the appropriate response to give to these questions would be to enumerate a list of open-and-shut criteria. If we were to attempt that, my prediction is these criteria would reflect the differentiae of true images and of imagination which I have offered above, in the analytic of imagination and images. These different functions and roles are in practice the indicators by which we are able to detect the presence of imagination and images that are valid. But they establish themselves empirically, by leading us to repentance and renewal, by sustaining faith, by evoking in us the responses of thanksgiving and adoration, by enabling us to build up a theological structure which interprets to our generation the insights given to another age and no longer immediately accessible for our contemporaries, and so on through the two lists. Phantasy is the process which attempts to do that but fails, because it has used a false or a debased currency, the image that broke in its hands. What we dare not think is that somehow we have a choice — to use or not to use the imagination and its media, images, in religion and in theology. Whether we acknowledge it or not, we have been employing imagination in our religion and in our theology, ever since we first became involved in these practices. It is a question, then, not of whether we employ it or not, but of how good, how irreproachable we can, by the grace of God, make our employment of it.